Disclaimer

The publisher of this book is by no way associated with the National Institute of Standards and Technology (NIST). The NIST did not publish this book. It was published by 50 page publications under the public domain license.

50 Page Publications.

Book Title: A Guide to Printed and Electronic Resources for Developing a Cost-Effective Risk Mitigation Plan for New and Existing Constructed Facilities

Book Author: Robert E. Chapman; Douglas S. Thomas

Book Abstract: Economic tools are needed to help the owners, managers, and designers of constructed facilities to select cost-effective combinations of mitigation strategies that respond to natural and man-made hazards. Economic tools include evaluation methods, standards that support and guide the application of those methods, and software for implementing the evaluation methods. Developing a cost-effective risk mitigation plan involves assessing the risks associated with natural and man-made hazards, formulating combinations of mitigation strategies for constructed facilities exposed to those hazards, and using economic tools to identify the most cost-effective combination of strategies. Developing a risk mitigation plan requires both guidance and data. Guidance is needed to help owners and managers to assess the risks facing their facility. Data about the frequency and consequences of natural and man-made hazards are needed when assessing the risks that a particular facility faces from these hazards. Estimates of the costs of protection are needed to insure that safeguarding personnel and physical assets and satisfying financial constraints are kept in balance. Although there is a great deal of high-quality information available on risk assessment and risk management, natural and man-made hazards, and economic tools, there is no central source of data and tools to which the owners and managers of constructed facilities and other key decision makers can turn for help in developing a cost-effective risk mitigation plan. This document provides an annotated bibliography of printed and electronic resources that serves as a central source of data and tools to help the owners, managers, and designers of constructed facilities develop a cost-effective risk mitigation plan.

Citation: NIST Interagency/Internal Report (NISTIR) - 7390

Keyword: building economics;construction;economic analysis;hazards;homeland security;risk assessment;terrorism

NISTIR 7390

 U.S. Department of Commerce
Technology Administration
National Institute of Standards and Technology

Office of Applied Economics
Building and Fire Research Laboratory
Gaithersburg, MD 20899

A Guide to Printed and Electronic Resources for Developing a Cost-Effective Risk Mitigation Plan for New and Existing Constructed Facilities

Robert E. Chapman and Douglas S. Thomas

NISTIR 7390

U.S. Department of Commerce
Technology Administration
National Institute of Standards and Technology

Office of Applied Economics
Building and Fire Research Laboratory
Gaithersburg, MD 20899

A Guide to Printed and Electronic Resources for Developing a Cost-Effective Risk Mitigation Plan for New and Existing Constructed Facilities

Robert E. Chapman and Douglas S. Thomas

Sponsored by:
National Institute of Standards and Technology
Building and Fire Research Laboratory

February 2007

U.S. DEPARTMENT OF COMMERCE
Carlos M. Gutierrez, Secretary

TECHNOLOGY ADMINISTRATION
Robert C. Cresanti, Under Secretary for Technology

NATIONAL INSTITUTE OF STANDARDS AND TECHNOLOGY
William A. Jeffrey, Director

Abstract

Developing a cost-effective risk mitigation plan involves assessing the risks associated with natural and man-made hazards, formulating combinations of mitigation strategies for constructed facilities exposed to those hazards, and using economic tools to identify the most cost-effective combination of strategies. Developing a risk mitigation plan requires both guidance and data. Guidance is needed to help owners and managers to assess the risks facing their facility. Data about the frequency and consequences of natural and man-made hazards are needed when assessing the risks that a particular facility faces from these hazards. Estimates of the costs of protection are needed to ensure that safeguarding personnel and physical assets and satisfying financial constraints are kept in balance. Finally, guidance on the use of economic evaluation methods is needed to ensure that the correct method, or combination of methods, is used. Although there is a great deal of high-quality information available on risk assessment and risk management, natural and man-made hazards, and economic tools, there is no central source of data and tools to which the owners and managers of constructed facilities and other key decision-makers can turn for help in developing a cost-effective risk mitigation plan. This document provides an annotated bibliography of printed and electronic resources that serves as that central source of data and tools to help the owners, managers, and designers of constructed facilities develop a cost-effective risk mitigation plan.

Keywords

Building economics; construction; economic analysis; hazards; homeland security; risk assessment; terrorism

Preface

This study was conducted by the Office of Applied Economics in the Building and Fire Research Laboratory at the National Institute of Standards and Technology. The study provides an annotated bibliography of printed and electronic resources that serves as a central source of data and tools to help the owners and managers of new and existing constructed facilities develop a cost-effective risk mitigation plan. The intended audience is the National Institute of Standards and Technology as well as other government and private sector organizations that are concerned with evaluating how to efficiently allocate scarce financial resources among security-related investment alternatives.

Disclaimer

Certain trade names and company products are mentioned in the text in order to adequately specify the technical procedures and equipment used. In no case does such identification imply recommendation or endorsement by the National Institute of Standards and Technology, nor does it imply that the products are necessarily the best available for the purpose.

Cover Photographs Credits

Looking west toward Biloxi from the east shore, many superstructure spans of US-90 Biloxi-Ocean Springs bridge were displaced north off their piers due to Hurricane Katrina (photo credit: J. O'Connor, Multidisciplinary Center for Earthquake Engineering Research).

Acknowledgements

The authors wish to thank all those who contributed so many excellent ideas and suggestions for this report. They include Dr. William Grosshandler of the Fire Research Division in the Building and Fire Research Laboratory (BFRL) at the National Institute of Standards and Technology (NIST), manager of BFRL's Research and Development for the Safety of Threatened Buildings Program, for his technical guidance, suggestions, and support. Special appreciation is extended to Dr. Harold E. Marshall and Dr. David Butry of BFRL's Office of Applied Economics (OAE) for their thorough reviews and many insights and to Ms. Tessa Beavers for her assistance in preparing the manuscript for review and publication. Special thanks are due to each of the OAE Technical Working Group members—Ms. Janet S. Baum, Health, Education + Research Associates, Inc.; Dr. Saul I. Gass and Dr. Howard K. Hung of NIST's Information Technology Laboratory; Mr. Robert N. Harvey, Washington Group Infrastructure Corporation; Mr. David Henry, U.S. Department of Commerce; Mr. Muthiah Kasi, Alfred Benesch & Company; Ms. Milagros Kennett, Federal Emergency Management Agency, U.S. Department of Homeland Security; Mr. Douglas N. Mitten, Project Management Services, Inc.; and Dr. Stephen R. Thomas, Construction Industry Institute—for their guidance throughout the development and production of this report. The report has also benefited from the review and technical comments provided by Mr. Stephen A. Cauffman of BFRL's Materials and Construction Research Division and Dr. M. Hayden Brown of BFRL's Office of Applied Economics.

Table of Contents

ABSTRACT V

PREFACE VII

ACKNOWLEDGEMENTS IX

1 INTRODUCTION 1
    1.1   BACKGROUND 1
    1.2   PURPOSE 2
    1.3   SCOPE AND APPROACH 3
    1.4   ORGANIZATION OF THE ANNOTATED BIBLIOGRAPHY 5

2 RISK ASSESSMENT RESOURCES 7
    2.1   RISK ASSESSMENT GUIDANCE DOCUMENTS 8
    2.2   RISK ASSESSMENT SOFTWARE 15
    2.3   HAZARDS DATA 20
        2.3.1 Natural Hazards 20
        2.3.2 Man-Made Hazards 27

3 RISK MANAGEMENT RESOURCES 33
    3.1   RISK MANAGEMENT GUIDANCE DOCUMENTS 33
    3.2   RISK MANAGEMENT SOFTWARE 55
    3.3   GUIDANCE DOCUMENTS FOR ESTIMATING COSTS AND LOSSES 57
        3.3.1 Mitigation Costs 57
        3.3.2 Event-Related Losses 65
    3.4   SOFTWARE FOR ESTIMATING COSTS AND LOSSES 71

4 ECONOMIC EVALUATION GUIDANCE 75
    4.1   ECONOMIC TOOLS 76
        4.1.1 Evaluation Methods 76
        4.1.2 Industry Standards 82
        4.1.3 Software for Implementing Industry Standards 90
    4.2   ECONOMIC MODELING RESOURCES 91
    4.3   ANALYSIS STRATEGIES FOR TREATING UNCERTAINTY 96

5 SUMMARY AND RECOMMENDATIONS FOR FURTHER RESEARCH 99
    5.1   SUMMARY 99
    5.2   RECOMMENDATIONS FOR FURTHER RESEARCH 99

APPENDIX A: OVERVIEW OF THE THREE-STEP PROTOCOL FOR DEVELOPING A COST-EFFECTIVE RISK MITIGATION PLAN 101
    A.1   PERFORM RISK ASSESSMENT 104
        A.1.1 Establish Risk Mitigation Objectives and Constraints 105
        A.1.2 Conduct Assessment and Document Findings 105

A.2   SPECIFY COMBINATIONS OF RISK MITIGATION STRATEGIES FOR EVALUATION 106
  A.2.1 Review Alternative Risk Mitigation Strategies 106
  A.2.2 Select Candidate Combinations of Risk Mitigation Strategies 109
  A.2.3 Develop Cost/Loss Estimates and Sequence of Cash Flows for Each Candidate Combination 109
A.3   PERFORM ECONOMIC EVALUATION 110
  A.3.1 Select Appropriate Economic Method(s) for Evaluating the Candidate Combinations of Risk Mitigation Strategies 111
  A.3.2 Compute Measures of Economic Performance for Each Candidate Combination 111
  A.3.3 Recompute Measures of Economic Performance Taking into Consideration Uncertainty 112
  A.3.4 Analyze Results and Identify the Most Cost-Effective Combination of Risk Mitigation Strategies 113
  A.3.5 Prepare Report with Documentation Supporting Recommended Risk Mitigation Plan 114

APPENDIX B: CLEARINGHOUSES AND WEB PORTALS 117

APPENDIX C: POLICIES, RESEARCH, AND THEORY 123

GLOSSARY OF RELATED TERMS 133

AUTHOR INDEX BY REFERENCE NUMBER 139

SUBJECT INDEX BY REFERENCE NUMBER 143

Acronyms and Abbreviations

AACE	Association for the Advancement of Cost Engineering
AASHTO	American Association of State Highway and Transportation Officials
ACECenter	Assessment of Catastrophic Events Center
AEO	Annual Energy Outlook
AER	Annual Energy Review
AHP	Analytical Hierarchy Process
AIR	Annual Interest Rate
ALFRED	ArchivaL Federal Reserve Economic Database
ALOFT-FT	A Large Outdoor Fire plume Trajectory model - Flat Terrain
ALOHA	Aerial Locations of Hazardous Atmospheres
AMA	American Management Association
AMMTIAC	Advanced Materials, Manufacturing, and Testing Information Analysis Center
AMPTIAC	Advanced Materials and Processes Technology Information Analysis Center
ANSI	American National Standards Institute
ANSS	Advanced National Seismic System
ASCE	American Society of Civil Engineers
ASCOS	Analysis of Smoke Control Systems
ASIS	American Society for Industrial Security
ASME	American Society of Mechanical Engineers
BEA	Bureau of Economic Analysis
BCR	Benefit-to-Cost Ratio
BEES	Building for Environmental and Economic Sustainability
BFRL	Building and Fire Research Laboratory
BLCC	Building Life-Cycle Cost
BLS	Bureau of Labor Statistics
BMSP	Blast Mitigation Structures Program
BTS	Bureau of Transport Statistics
CAD	Computer Aided Design
CARVER	Criticality, Accessibility, Recoverability, Vulnerability, Effect, and Recognizability
CATS	Consequences Assessment Tool Set
CBECS	Commercial Buildings Energy Consumption Survey
CBR	Chemical Biological and Radiological
CBS	Central Broadcasting System
CCE	Certified Cost Engineer
CCTV	Closed Circuit Television
CESB	Council of Engineering and Scientific Specialty Boards
CONTAM	Contaminant Multizone Modeling Software
CORE	Coastal and Ocean Resource Economics
CPP	Certified Protection Professional

CPSC	Consumer Product Safety Commission
CREATE	Center for Risk and Economic Analysis of Terrorism Events
CSB	Chemical Safety Board
CSI	Construction Specifications Institute
DARP	Damage Assessment and Restoration Program
DCSINT	Deputy Chief of Staff for Intelligence
DGN	Microstation Design file (file format)
DHS	Department of Homeland Security
DOD	Department of Defense
DOE	Department of Energy
DOJ	Department of Justice
DOT	Department of Transportation
DTRA	Defense Threat Reduction Agency
DWG	AutoCAD drawing (file format)
ECES	Environmental Cost Element Structure
ECHOS	Environmental Cost Handling Options and Solutions
EERC	Energy Escalation Rate Calculator
EERE	Energy Efficiency and Renewable Energy
EIA	Energy Information Administration
ENR	Engineering News Record
EPA	Environmental Protection Agency
EPP	Environmentally Preferable Purchasing
EqIP	Earthquake Information Providers Group
EQNET	EarthQuake Information NETWork
ESI	Estimating Systems Incorporated
ESRI	Environmental Systems Research Institute
FBI	Federal Bureau of Investigation
FEMA	Federal Emergency Management Agency
FEMP	Federal Energy Management Program
FHWA	Federal Highway Administration
FIRST	FIRe Simulation Technique
FLASH	Federal Alliance for Safe Homes
FRED	Federal Reserve Economic Database
GAO	Government Accountability Office
GIS	Geographic Information System
GPO	Government Printing Office
GSA	General Services Administration
HAZUS	Hazards U.S.
HAZUS-MH	HAZUS Multi-Hazard
HPAC	Hazard Prediction and Assessment Capability
HSDL	Homeland Security Digital Library
HSPD	Homeland Security Presidential Directives
HVAC	Heating Ventilation and Air Conditioning
IBHS	Institute for Business & Home Safety
ICBO	International Conference of Building Officials
ICC	International Code Council

ICE	Integrated Computer Engineering
IOC	Intergovernmental Oceanographic Commission
ISBE	Infrastructure Security for the Built Environment
ISDR	International Strategy for Disaster Reduction
ITA	International Trade Administration
ITI	Innovative Technologies Institute
ITIC	International Tsunami Information Center
LCC	Life Cycle Cost
MADA	Multi-Attribute Decision Analysis
MAE	Mid-America Earthquake Center
MIPT	Memorial Institute for the Prevention of Terrorism
NASA	National Aeronautics and Space Administration
NBER	National Bureau of Economic Research
NCDC	National Climatic Data Center
NEHRP	National Earthquake Hazards Reduction Program
NEIC	National Earthquake Information Center
NEMA	National Emergency Management Association
NESDIS	National Environmental Satellite, Data, and Information Service
NFIP	National Flood Insurance Program
NFIRS	National Fire Incident Reporting System
NFPA	National Fire Protection Association
NIBS	National Institute of Building Sciences
NIMS	National Incident Management System
NIPP	National Infrastructure Protection Plan
NIST	National Institute of Standards and Technology
NOAA	National Oceanic and Atmospheric Association
NODC	National Oceanographic Data Center
NSF	National Science Foundation
NWS	National Weather Service
OAE	Office of Applied Economics
OLES	Office of Law Enforcement Standards
OSHA	Occupational Safety and Health Administration
PATH	Partnership for Advancing Technology in Housing
PBS	Public Buildings Service
PCI	Professional Certified Investigator
PDC	Pacific Disaster Center
PERI	Public Entity Risk Institute
PML	Probable Maximum Loss
PSP	Physical Security Professional
PSHA	Probabilistic Seismic Hazard Analysis
RAM	Risk Assessment Methodology
RAMCAP	Risk Analysis and Management for Critical Asset Protection
RAMPART	Risk Assessment Method – Property Analysis and Ranking Tool from the General Services Administration
RMS	Risk Management Solutions
ROI	Return on Investment

SBCCI	Southern Building Code Congress International, Inc.
SITC	Swiss Insurance Trainee Center
SL	Scenario Loss
START	Study of Terrorism and Responses to Terrorism
TCO	Total Cost of Ownership
TISP	The Infrastructure Security Partnership
TRI	Toxic Release Inventory
TSR	Tropical Storm Risk
USACE	United States Army Corps of Engineers
USFA	United States Fire Administration
USGS	United States Geological Survey
UTL	Universal Task List
VA	Value Analysis
VAPO	Vulnerability Assessment and Protection Option
ZIP	Zone Improvement Plan

1 Introduction

1.1 Background

The September 11, 2001 attacks on the World Trade Center and the Pentagon, and the subsequent dispersion of anthrax through the postal system, changed the way many in the United States approach security and safety. These events have prompted the owners and managers of constructed facilities—buildings, industrial facilities, and other physical infrastructure—to address terrorist risks and protect the occupants, property, and functions of their facilities.

The devastation to the Gulf Coast caused by Hurricanes Katrina and Rita, and their impact on the national economy, underscored the need to plan for natural and man-made disasters as well as terrorist threats. Each year, natural and man-made disasters cause on average an estimated $52 billion in damages in the United States in terms of lives lost, disruption of commerce, properties destroyed, and the costs of mobilizing emergency response personnel and equipment.[1] It is important, however, to note that a single major disaster, such as Hurricane Katrina, can result in damages in the United States that greatly exceed that average figure.

These realities have led to changes in how key decision-makers respond to natural and man-made hazards. Among these changes are the way owners and managers think about the design, location, construction, operation, and renovation of constructed facilities. The range of responses available to decision-makers is extensive, as is the potential expense. Parallel to the reality of the risks posed by natural and man-made hazards is the reality of budget constraints. Owners and managers of constructed facilities are confronted with the challenge of planning for and responding to natural and man-made hazards in a financially responsible manner. The two objectives—safeguarding personnel and physical assets and satisfying financial constraints—must be balanced through a cost-effective risk mitigation plan.

Emerging from this new focus on planning is the realization that it makes sense to evaluate all kinds of natural and man-made hazards as a group. Costs for protection against multiple hazards can be shared among the hazards protected against, thereby reducing the cost of any single form of protection. Or, looked at in another way, a given cost of protection can yield extra benefits when considering multiple hazards. This spillover of benefits from one kind of protection to another highlights the need for a holistic approach to planning protection against multiple hazards.

---

[1] National Science and Technology Council, Committee on Environmental and Natural Resources, Subcommittee on Disaster Reduction, *Grand Challenges for Disaster Reduction,* June 2005, p. 3.

In two earlier studies,[2, 3] the Office of Applied Economics (OAE) in the Building and Fire Research Laboratory (BFRL) at the National Institute of Standards and Technology outlined a three-step protocol for developing a cost-effective risk mitigation plan. The protocol has been revised and expanded to meet the need for a holistic approach to planning protection against multiple hazards. In January 2006, the protocol was submitted to ASTM International for promulgation as a Standard Guide for Developing a Cost-Effective Risk Mitigation Plan; it was approved by ASTM and released as Standard Guide E2506 in October 2006.[4] The three-step protocol described in Standard Guide E2506 helps decision-makers (1) assess the likelihood that their facility and its contents will be damaged from natural and man-made hazards; (2) identify engineering, management, and financial strategies for abating the risk of damages; and (3) use standardized economic evaluation methods to select the most cost-effective combination of risk mitigation strategies to protect their facility. The three-step protocol is the basis for our approach in this report. Appendix A provides an overview of the protocol and links the steps in the protocol to the three main chapters in this report.

Developing a cost-effective risk mitigation plan requires both guidance and data. Guidance is needed to help owners and managers to assess the risks facing their facility or collection of facilities. Data about the frequency and consequences of natural and man-made hazards are needed when assessing the risks that a particular facility faces from these hazards. Estimates of the costs of protection are needed to ensure that safeguarding personnel and physical assets and satisfying financial constraints are kept in balance. Finally, guidance on the use of economic evaluation methods is needed to ensure that the correct method, or combination of methods, is used.

1.2 Purpose

Research by the authors has shown that, although there is a great deal of high-quality information available on risk assessment and risk management, natural and man-made hazards, and methods for measuring economic performance, there is no central source of data and tools to which the owners and managers of constructed facilities and other key decision-makers can turn for help in developing a cost-effective risk mitigation plan. The purpose of this document is to provide an annotated bibliography of printed and electronic resources that serves as that central source of data and tools to help readers develop a cost-effective risk mitigation plan.

This report is being published both as a printed document and as a web-based publication to facilitate its widespread use. It is intended as a ready reference for researchers, public and private sector decision-makers, and practitioners—owners and managers of

---

[2] Chapman, Robert E. and Leng, Chi J. *Cost-Effective Responses to Terrorist Risks in Constructed Facilities*, NISTIR 7073 (Gaithersburg, MD: National Institute of Standards and Technology, March 2004).
[3] Marshall, Harold E., Chapman, Robert E., and Leng, Chi J. "Risk Mitigation Plan for Optimizing Protection of Constructed Facilities," *Cost Engineering*, Vol. 46, No. 8, August 2004, pp. 26-33.
[4] ASTM International. "Standard Guide for Developing a Cost-Effective Risk Mitigation Plan for New and Exixting Constructed Facilities," E2506. Annual Book of ASTM Standards: 2007. Vol.4.12. West Conshohocken, PA: ASTM International.

constructed facilities, architects, engineers, constructors, and other providers of professional services for constructed facilities—concerned with cost-effective responses to natural and man-made hazards. It is also intended for practitioners interested in formulating and evaluating alternative combinations of risk mitigation strategies.

The key difference between the two publication media is that the printed document is index-based. The Subject Index and Author Index point the reader to key reference documents, databases, and software tools that will help them develop a cost-effective risk mitigation plan. In the web-based version posted on the BFRL web site, all web links are active, enabling practitioners to browse documents and data sources. Furthermore, many of the web links permit documents and data files to be downloaded for future reference and use. Given that new security-related materials are being published on a regular basis, this report (both printed and web-based) is planned for periodic revision to capture new source documents and web links.

1.3 Scope and Approach

This report covers in a limited manner the key steps for developing a cost-effective risk mitigation plan. The primary emphasis in the report, however, is placed on summarizing key reference documents and data sources. Our intent is to provide readers with sufficient guidance that they can readily access documents, databases, and other key information resources in support of their efforts to develop a cost-effective risk mitigation plan for addressing multiple hazards. The document establishes a framework for developing a cost-effective risk mitigation plan in which each component complements and reinforces the others.

Because constructed facilities face a wide range of hazards—both natural and man-made—this report identifies information resources for each type of hazard. A major emphasis is placed on providing information on five natural hazards—earthquakes, hurricanes, tornadoes, wildfires, and coastal and river flooding—and three man-made hazards—terrorism, other criminal acts, and chemical accidents. A limited amount of information on three additional natural hazards—volcanic eruptions, landslides, and tsunamis—is also provided.

This report contains four chapters and three appendices in addition to the Introduction. Chapters 2 through 4, the major contributions of this report, provide background information and an annotated bibliography covering major reference and guidance documents and data sources that will help the reader formulate a cost-effective risk mitigation plan. The background and annotated bibliography sections of each chapter are designed to give readers a good grounding in each section's topic and reference material. Our goal is to give the reader enough guidance and reference material that they can quickly form an action plan for developing cost-effective responses to multiple hazards.

Chapter 2 provides references to risk assessment resources. The chapter is divided into three sections. Section 2.1 focuses on risk assessment guidance documents. The emphasis in Section 2.1 is on establishing a technical foundation for conducting a risk assessment and identifying key resource documents that provide guidance on how to conduct a multi-hazard risk assessment. Section 2.2 focuses on risk assessment software. Software tools are included because they are of great value in compiling, organizing, and evaluating information and data needed to conduct the risk assessment. Section 2.3 covers sources of hazards data; it is divided into two subsections: natural hazards and man-made hazards. Each subsection provides descriptive material covering key data sources. Our intent is to provide readers with sufficient background information that they can decide where to go to get additional data to support their risk assessment evaluations.

Chapter 3 provides references to risk management resources. The chapter is divided into four sections. Section 3.1 focuses on risk management guidance documents. The emphasis in Section 3.1 is on establishing a technical foundation for formulating combinations of risk mitigation strategies that address the hazards identified in the risk assessment and identifying key resource documents that provide guidance on how to formulate multi-hazard risk mitigation strategies. Section 3.2 focuses on risk management software. Software tools are included because they are of great value in compiling, organizing, and evaluating information and data needed to formulate alternative combinations of risk mitigation strategies. Section 3.3 covers guidance documents for cost-related issues; it is divided into two subsections: mitigation costs and event-related losses. Section 3.4 covers software tools that are useful in developing estimates of mitigation costs and event-related losses.

Chapter 4 provides references on how to conduct an economic evaluation. Section 4.1 provides an overview of economic tools. Economic tools include evaluation methods, standards that support and guide the use of those methods, and software for implementing the evaluation methods. The section also includes a discussion of how to choose the appropriate economic evaluation method or a combination of evaluation methods. Section 4.2 focuses on empirical considerations. Section 4.2 focuses on assumptions, the key parameters that define the project (e.g., purpose of the project, length of the study period, and discount rate), and the specification of alternative combinations of risk mitigation strategies. Section 4.3 describes an approach to help analysts treat uncertainty.

Chapter 5 concludes with a summary and a discussion of areas of need for future research.

Appendix A provides an overview of the three-step protocol. We provide the reader with an explanation of the three-step protocol to establish guiding principles for more effectively using the annotated bibliography presented in Chapters 2 through 4 to develop a cost-effective risk mitigation plan.

Appendix B provides links to clearinghouses and web portals where additional descriptive material and links to risk assessment/risk management guidance documents and software tools are provided.

Appendix C contains reference material on disaster-related policies and research. Summaries of theoretical studies on risk perception, the impacts of terrorism on the national economy, the challenge of quantifying terrorism risk for the insurance market, and other selected theoretical issues are provided.

1.4 Organization of the Annotated Bibliography

The annotated bibliography presented in Chapters 2 through 4 constitutes the majority of this report. Bibliographic references are provided within each section/subsection of Chapters 2 through 4. References are listed in alphabetical order by author, where the author may be a person, a company, an organization, or a government entity. Each reference is numbered sequentially and tied uniquely to the section/subsection in which it appears. A URL is provided whenever a reference is available in electronic format.[5] We have prepared an abstract for each reference that summarizes the salient points of the reference. In preparing each abstract, we examined each reference to better link it into the annotated bibliography. We believe this approach will help users find the materials they need to develop a cost-effective risk mitigation plan.

It is important to note that a reference may be relevant for more than one section/ subsection. To minimize duplication within this document, whenever a reference is relevant for more than one section/subsection, we provide the abstract in the most relevant section. All other listings of that reference refer the reader to the unique reference number of the section/subsection where the abstract appears. Several references are linked to other references. In these cases, the unique reference number for the linked reference is given in the abstract.

This document also contains an Author Index and a Subject Index. The Author Index is based on the first line of the reference, where the author is listed (e.g., American Society for Industrial Security). Each reference associated with that author is then listed. For example, one reference for the American Society for Industrial Security is 2.1.1. Turning to that reference, you will find the abstract and a link to an electronic version of the *General Security Risk Assessment Guideline*.

The Subject Index is organized as a two-tiered hierarchy. Each reference is classified using keywords drawn from the Subject Index. Elements in the first tier, such as Security, are listed first. Elements in the second tier are indented beneath the first tier element. References may be listed with a first tier element or with a second tier element. For example, Reference 2.1.1 is listed with the first tier element Security.

Several subjects refer you to another subject. These cases are highlighted with an *italics* font. For example, the generic term Mitigation instructs you to *See Individual Hazards*. Individual Hazards include but are not limited to Earthquakes, Floods, and Hurricanes.

---

[5] In most cases, references are available for free as downloads. In other cases, the URL provides information on how to purchase the reference.

2　Risk Assessment Resources

The first step in creating a cost-effective risk mitigation plan is a risk assessment for the facility or group of facilities to be protected. The risk assessment step is critical in generating the information needed to formulate the alternative combinations of risk mitigations strategies.

The risk assessment step includes specification of the decision-maker's objectives, the facilities to be protected, the natural and man-made hazards to be considered, the composition of the risk assessment team, and documentation procedures. The risk assessment involves data collection to establish the likelihood of natural and man-made hazards as well as the on-site collection and documentation of facility vulnerabilities to those hazards. Estimates of the value of the facility's assets and the consequences of an event occurring are also produced as part of the risk assessment.

Section 2.1 covers risk assessment guidance documents. The emphasis in Section 2.1 is on establishing a technical foundation for conducting a risk assessment and identifying key resource documents that provide guidance on how to conduct a multi-hazard risk assessment. Documents abstracted include a series of FEMA *How-to Guides*, as well as professional society publications from ASCE, ASIS, ASME, ASTM, and NFPA.

Section 2.2 covers risk assessment software. Coverage of software tools focuses on their value in compiling, organizing, and evaluating information and data needed to conduct the risk assessment. Software tools from both the public and private sectors are abstracted. Public sector tools include HAZUS-MH, which covers earthquakes, wind, and floods, and the Risk Assessment Database, which can be used for both natural and man-made hazards; both software tools were developed with funding from FEMA. Private sector tools cover a wide variety of hazards (natural and man-made), simulation-modeling capabilities useful in performing a probabilistic risk assessment, and the use of geographical information system (GIS) products.

Section 2.3 covers sources of hazards data; it is divided into two subsections: natural hazards and man-made hazards. Each subsection provides descriptive material covering key data sources. Historical patterns of natural disasters indicate which areas are more prone to these specific hazards in the future. The Federal Government maintains a wide variety of hazard-specific databases covering earthquakes, floods, hurricanes, tornadoes, and wildfires. Significant Federal Government resources abstracted include databases maintained by FEMA, NASA, NOAA, NWS, and the USGS. Databases covering man-made hazards include detailed information on crime and environmental disasters. Information on terrorism is also presented. Non-federal data sources covering both natural and man-made hazards include NFPA, the insurance industry, and a number of private sector organizations. The goal of the abstracted data sources is to provide readers with sufficient background information that they can decide where to go to get additional data to support their risk assessment evaluations.

2.1 Risk Assessment Guidance Documents

2.1.1 American Society for Industrial Security. General Security Risk Assessment Guideline. Alexandria, VA: ASIS International, 2003.
URL: http://www.asisonline.org/guidelines/guidelinesgsra.pdf

This ASIS publication provides a method to identify security risks at a specific location using a seven-step process: (1) Understand the organization and identify the people and assets at risk; (2) Specify loss risk events/vulnerabilities; (3) Establish the probability of loss risk and frequency of events; (4) Determine the impact of the event; (5) Develop options to mitigate risks; (6) Study the feasibility of implementation of options; and (7) Perform a cost/benefit analysis. Each step is detailed in the publication. It also defines key terms, designates guidelines, and provides information sources.

2.1.2 ASTM International. "Standard Guide for the Estimation of Building Damageability in Earthquakes." E 2026. Annual Book of ASTM Standards: 2005. Vol. 04.12. West Conshohocken, PA: ASTM International.
URL: http://www.astm.org/

This guide presents specific approaches to estimate "probable maximum loss" (PML), scenario loss (SL), and establishes a commercial standard for studying expected loss in the event of an earthquake. It discusses loss estimation, probability, seismic risk, computer assessment tools, and building stability. It is intended to be used by the real estate and technical communities to evaluate the vulnerability of buildings in an earthquake.

2.1.3 American Society of Civil Engineers. Seismic Evaluation of Existing Buildings. SEI/ASCE 31-03. Reston, VA: American Society of Civil Engineers, 2003.

This standard provides a "three-tiered process" to evaluate a building's performance in the event of seismic activity. It is intended to replace FEMA 310, "Handbook for the Seismic Evaluation of Buildings," and be used by design professionals, code officials, and building owners. It is compatible with FEMA 356, "Pre-standard and Commentary for the Seismic Rehabilitation of Buildings."

2.1.4 American Society of Mechanical Engineers. Risk Analysis and Management for Critical Asset Protection for Terrorist Threats and Homeland Security. Washington DC: ASME-ITI, 2005.
URL: http://www.asme-iti.org/RAMCAP/RAMCAP.cfm

The American Society of Mechanical Engineers (ASME) Innovative Technologies Institute (ITI), through funding from the Department of Homeland Security (DHS), has launched the Risk Analysis and Management for Critical Asset Protection (RAMCAP) project. In 2005, ASME-ITI published a guidance document on assessing the risk associated with terrorist threats. The goal of this document is to inform resource allocation decisions for the protection of critical infrastructure. Although the focus is on terrorist threats, the ASME-ITI guidance document provides a framework suitable for addressing other types of man-made hazards as well as natural hazards. Specifically, the document provides a review of the existing approaches to assessing risk, highlights the common terminology and basis for reporting results, and presents recommended methodology and best practices.

2.1.5 Deisler, Paul F., Jr. "A Perspective: Risk Analysis as a Tool for Reducing the Risks of Terrorism." Risk Analysis. 22 (2002): 405-413.

This article discusses the extent that risk analysis can be used to address the issue of terrorism. Each incident of terrorism has unique characteristics, but they also have common traits. This article is part of a collection of works discussed in reference 3.3.2.18.

2.1.6 Federal Emergency Management Agency.
URL: http://www.fema.gov/

The Federal Emergency Management Agency (FEMA) is part of the Department of Homeland Security and is the primary organization responsible for mitigating, preparing, and responding to major disasters in the United States. FEMA provides a free hazard identification program called HAZUS, which is a useful tool to estimate natural hazard losses before and after a disaster. FEMA also provides online hazard mapping, which is searchable by location (address or latitude and longitude) and the type of hazard (e.g., hurricane or tornado). Access to historic disaster maps, statistical information, and disaster statistics is available on the FEMA website and is free to the public. The National Flood Insurance Program (NFIP) is managed by FEMA and was established to identify and reduce the increasing costs of flood damage (http://www.floodsmart.gov). Flood insurance maps and flood insurance study reports organized by area are available both on hard copy and digital media through NFIP. Flood statistics are available online through the NFIP at http://www.fema.gov/business/nfip/statistics/pcstat.shtm. For fire statistics visit the United States Fire Administration (USFA). It is organized through FEMA with the purpose of minimizing damage and suffering caused by fire disasters (http://www.usfa.fema.gov/). FEMA's website contains information on the following topics: dam failure, earthquakes, fires, floods, hazardous materials, chemicals, heat, hurricanes, landslides, nuclear issues, terrorism, thunderstorms, tornados, tsunamis, and general hazards.

2.1.7 Federal Emergency Management Agency. "Mitigation Planning Resources."
URL: http://www.fema.gov/plan/mitplanning/planning_resources.shtm

The Federal Emergency Management Agency (FEMA) has developed a series of guidance manuals to assist state and local communities in planning for risk mitigation. These manuals address the need for risk assessment for a variety of hazards. They describe the processes of identifying hazards, identifying and developing mitigation strategies, implementing risk mitigation plans, and applying these processes to man-made hazards. All of the FEMA guidance manuals are designed to be used at the community level rather than at the level of individual businesses or buildings. But building owners and managers may benefit from increased awareness of local hazards and the types of personnel and expertise that FEMA recommends, particularly if they undertake risk mitigation in a coordinated fashion with local emergency responders.

2.1.8 Federal Emergency Management Agency. How-to Guide: Using HAZUS-MH for Risk Assessment. FEMA 433. Washington, DC: Federal Emergency Management Agency, August 2004.
URL: http://www.fema.gov/plan/prevent/rms/rmsp433.shtm

This How-to Guide aids users of HAZUS-MH (see reference 2.2.7 on page 16) in preparing sound risk assessments. It is targeted toward those who have some familiarity with the HAZUS-MH software and is interested in using the software to support risk assessment studies. There are five steps in this guide: (1) Identify Hazards; (2) Profile Hazards; (3) Inventory Assets; (4) Estimate Losses; and (5) Consider Mitigation Options. This text takes the reader through the complete process of risk assessment using HAZUS-MH.

2.1.9 Federal Emergency Management Agency. How-to Guide #1: Getting Started: Building Support for Mitigation Planning. FEMA 386-1. Washington, DC: Federal Emergency Management Agency, September 2002.
URL: http://www.fema.gov/plan/mitplanning/howto1.shtm

FEMA How to Guide #1: *Getting Started: Building Support for Mitigation Planning* directs public and private organizations in the initial mitigation planning stages. This how-to guide proposes a three-step process for the initial planning stage. Step one is to assess community support. For this step the guide provides methods of identifying knowledge, support, and resources available to the area. Step two consists of building a planning team. It is an overview of appointing individuals who will be responsible for seeing the planning process through. The final step is to engage the public. The guide assists in identifying stakeholders, organizing public participation, and bringing public feedback.

2.1.10 Federal Emergency Management Agency. How-to Guide #2: Understanding Your Risks. FEMA 386-2. Washington, DC: Federal Emergency Management Agency, September 2002.
URL: http://www.fema.gov/plan/mitplanning/howto2.shtm

FEMA How-To Guide #2: *Understanding Your Risks* addresses natural hazards and offers descriptions of the risk assessment process that can be generalized to other types of hazards. The four-step process consists of: (1) identifying the hazards; (2) profiling the hazard events to determine magnitudes and pinpoint more specific asset vulnerabilities; (3) inventorying assets; and (4) estimating losses. Guide #2 is very helpful in identifying the risks of a specific geographic area. It identifies sources of information, methods of evaluation, asset inventory processes, and loss estimation methods.

2.1.11 Federal Emergency Management Agency. How-to Guide #3: Developing the Mitigation Plan. FEMA 386-3. Washington, DC: Federal Emergency Management Agency, September 2002.
URL: http://www.fema.gov/plan/mitplanning/howto3.shtm

How-To Guide #3: *Developing the Mitigation Plan* provides state and local decision-makers with the tools to identify mitigation objectives and strategies. It provides a four-step process to reduce or eliminate the loss of life and property in the event of a disaster. Step one is to develop mitigation goals and objectives. For this step, the guide provides a general overview of goals and objectives to risk mitigation. Step two is to identify and prioritize mitigation actions. The third step is to prepare an implementation strategy. It provides methods to implement the actions identified in step two. The final step is to document the mitigation planning process, which is detailed in the guide.

2.1.12 Federal Emergency Management Agency. How-to Guide #4: Bringing the Plan to Life. FEMA 386-4. Washington, DC: Federal Emergency Management Agency, September 2002.
URL: http://www.fema.gov/plan/mitplanning/howto4.shtm

How-To Guide #4: *Bringing the Plan to Life* describes the steps that planners can take to implement the strategies, which were identified in How-To Guide #3, to accomplish the stated risk mitigation objectives. This process includes four steps: (1) adopt the mitigation plan; (2) implement the plan recommendations; (3) evaluate your planning results; (4) and revise the plan. Guide #4 contains the implementation process that identifies the necessary actions to establish and maintain an effective risk reduction plan.

2.1.13 Federal Emergency Management Agency. How-to Guide #6: Integrating Historic Property and Cultural Resource Considerations into Hazard Mitigation Planning. FEMA 386-6. Washington, DC: Federal Emergency Management Agency, September 2002.
URL: http://www.fema.gov/plan/mitplanning/howto6.shtm

How-To Guide #6: *Integrating Historic Property and Cultural Resource Considerations into Hazard Mitigation Planning* addresses the preservation of historic sites and artifacts in the event of a disaster. Since historic sites and artifacts are irreplaceable and are many times an integral part of the community and economy, it is important to consider their loss or possible damage in the event of a disaster. How-to Guide #6 proposes a four-phase process to evaluate the risk to these cultural resources. The first phase, Organize Resources, consists of assessing community support, building a planning team, and engaging the public. Phase two, Assess Risks, is comprised of four steps: identify hazards, profile hazard events, inventory assets, and estimate losses. The third phase, Develop a Mitigation Plan, identifies mitigation actions and implementation strategies for protecting identified historic property. The final phase is to implement the plan, which consists of a five-step process.

2.1.14 Federal Emergency Management Agency. How-to Guide #7: Integrating Manmade Hazards. FEMA 386-7. Washington, DC: Federal Emergency Management Agency, September 2002.
URL: http://www.fema.gov/plan/mitplanning/howto7.shtm

How-To Guide #7: *Integrating Manmade Hazards* directly relates to terrorism and "technological disasters." Because of the attack on the World Trade Center, the Okalahoma City bombing, and other terrorist attacks the importance of preparing for a manmade disaster has become a more serious component of risk mitigation. Guide #7 proposes a four-phase approach to mitigate manmade hazards. The first phase, organize resources, pulls resources together to get a mitigation plan off the ground. Phase 2 of Guide #7, assess risk, involves identifying hazards and estimating the possible loss in the event of a disaster. Developing goals, prioritizing actions, preparing the strategy, and documenting the process are outlined in phase three, which is to develop a mitigation plan. The final phase is to implement the plan. The four phases of How-To Guide #7 help reduce the vulnerability to future attacks.

2.1.15 Federal Emergency Management Agency. Risk Management Series Publications.
URL: http://www.fema.gov/plan/prevent/rms/index.shtm

See reference Appendix B:7 (page 119) and the *Risk Management Series Publications* that follow it

2.1.16 Hahn, Robert W., and Anne Layne-Farrar. The Law and Economics of Software Security. Working Paper 06-08. Joint Center: AEI-Brookings Joint Center for Regulatory Studies, April 2006.
URL: http://aei-brookings.org/admin/authorpdfs/page.php?id=1266

This working paper is to be printed in the Harvard Journal of Law and Public Policy. It provides an assessment of software security issues using a framework based on law and economics and discusses the current challenges of software security.

2.1.17 Kunreuther, Howard, Robert Meyer, and Christophe Van den Bulte. Risk Analysis for Extreme Events: Economic Incentives for Reducing Future Losses. NIST GCR 04-871. Gaithersburg, MD: National Institute of Standards and Technology, October 2004.
URL: http://www.bfrl.nist.gov/oae/publications/gcrs/04871.pdf

This NIST publication discusses risk assessment, risk perception, and risk management in relation to developing a strategy to deal with disasters. It details the difference between calculated risk and risk perception along with individual decision processes. Moreover, in this publication risk management is a key topic that addresses how mitigation, insurance, and building codes can reduce losses and provide funds for recovery. Three organizations provided information and research as part of this publication: RMS (see reference 2.1.22

on page 14), EQECAT (see reference 3.2.7 on page 57), and AIR (see reference 2.2.2 on page 15). RMS and EQECAT provided information to analyze earthquake mitigation, while AIR provided information to analyze hurricane hazards.

2.1.18 Leson, Joel. Assessing and Managing the Terrorism Threat. NCJ 210680 Washington DC: Bureau of Justice Assistance, September 2005.
URL: http://www.ncjrs.gov/pdffiles1/bja/210680.pdf

This publication of the Department of Justice outlines the components of risk assessment and management of terrorist threats. The process includes six assessments: critical infrastructure and key asset inventory, criticality assessment, threat assessment, vulnerability assessment, risk calculation, and countermeasure identification. This publication discusses each process and gives a relative scale to measure them.

2.1.19 National Fire Protection Association. Extreme Event Mitigation in Buildings. Quincy, MA: National Fire Protection Association, 2006.
URL: http://www.nfpa.org

This text provides guidance on designing and assessing buildings for performance and safety as it relates to natural and man-made disasters. It discusses evacuations, consequences, chemical and biological threats, and design strategies. *Extreme Event Mitigation in Buildings* is available for purchase from the National Fire Protection Association.

2.1.20 Public Entity Risk Institute.
URL: http://www.riskinstitute.org/

The Public Entity Risk Institute (PERI) is a not-for-profit organization founded in 1996 that disseminates risk management information to small public and private organizations. PERI is currently building a database of loss information, provides training in risk management, and is developing a metric for performance measurement.

2.1.21 RAND Corporation.
URL: http://www.rand.org/

RAND Corporation is a private nonprofit organization that conducts research on issues that concern society and government. These issues include the arts, child policy, civil justice, education, energy, health care, national security, terrorism, population, public safety, substance abuse, transportation, and the workforce. Much of the research at RAND is published in the RAND Review. A brief article in its fall 2005 issue discusses the Terrorism Risk Insurance Act, which was passed after the events of September 11. RAND has also been working with the Oklahoma City National Memorial Institute for the Prevention of Terrorism (MIPT) in building a database of terrorism incidents to help researchers, analysts, and others to prevent terrorism (http://www.rand.org/ise/projects/terrorismdatabase/index.html). This project includes the MIPT Terrorism Knowledge Base (http://www.tkb.org/Home.jsp), which has two databases: the RAND Terrorism Chronology Database and the RAND-MIPT Terrorism Incident Database. The former records international terrorist incidents that occurred between 1968 and 1997. The latter records domestic and international terrorist incidents occurring from 1998 to the present. The website includes a great deal of information about groups, incidents, cases, leaders, countries and other issues. See MIPT in reference 3.3.2.11

2.1.22 Risk Management Solutions Incorporated.
URL: http://www.rms.com/

Risk Management Solutions (RMS) provides products and services for quantifying and managing catastrophe risks. RMS conducts many types of risk modeling research: natural hazard, terrorism, weather, and enterprise risk. RMS produced its second version of RMS U.S. Terrorism Risk Model, which is a risk model for man-made disasters related to terrorism and weapons of mass destruction. Many of the models developed at RMS are used in the insurance market. A number of their publications are available free to the public, but some of their online services are password protected.

2.1.23 Swiss Re. May 8, 2006.
URL: http://www.swissre.com/

See reference 3.3.2.22 (page 69)

2.1.24 Willis, Henry H., Andrew R. Morral, Terrence K. Kelly, and Jamison Jo Medby. Estimating Terrorism Risk. Santa Monica, CA: RAND Corporation, 2005.
URL: http://www.rand.org/pubs/monographs/2005/RAND_MG388.pdf

This RAND publication reports on research conducted on the risk of terrorism. It discusses risk assessment, uncertainty, resource allocation, risk estimation, risk indicators, population based metrics, simulation models, event-based models, and aggregated risk estimators as they relate to terrorism. This document is available free to the public through the RAND Corporation (see reference 2.1.21 on page 13).

2.1.25 Woo, Gordon. "The Evolution of Terrorism Risk Modeling." Journal of Reinsurance. April 22, 2003.
URL: http://www.rms.com/Publications/EvolutionTerRiskMod_Woo_JournalRe.pdf

This article briefly discusses the different methods that have been used to model terrorism risk; it includes the deterministic scenario loss model, expert judgment, and terrorist target selection. This article explores basic philosophies in a non-technical manner.

2.1.26 Woo, Gordon. "Understanding Terrorism Risk." Risk Management Solutions. 2004.
URL: http://www.rms.com/Publications/UnderstandTerRisk_Woo_RiskReport04.pdf

This article published through Risk Management Solutions (see reference 2.1.22 on page 14) discusses quantitative approaches to understanding terrorism. It discusses terrorist trends, the use of game theory, steps to construct a risk curve, and terrorism insurance.

2.2 Risk Assessment Software

2.2.1 Air Dispersion Modeling Incorporated.
URL: http://www.air-dispersion-model.com/

Air Dispersion Modeling offers several air dispersion software programs; two programs have been used for research by the U.S. Environmental Protection Agency (EPA): CALPUFF and CALRoads. CALPUFF combines GIS and 3D models to create a complete outdoor air dispersion model. It has been used to model air quality scenarios related to toxic pollutant deposition, forest fire impacts, and visibility assessments. CALRoads is used to predict air quality impacts of pollutants near roadways. Air Dispersion has also developed an accidental toxic gas release model, SLAB View. It models releases from four types of sources: ground-level evaporating pool, elevated horizontal jet, elevated vertical jet, and ground-based instantaneous release. These software products are available for purchase on Air Dispersion Modeling Incorporated website.

2.2.2 AIR.
URL: http://www.air-worldwide.com

AIR provides analytical tools and software systems to manage natural and man-made disaster risks. Their products are primarily intended for insurers, reinsurers, and risk managers. The AIR Terrorism Loss Estimation Model estimates property and workers' compensation losses from a possible terrorist attack. Due to the uncertainty of terrorist attacks, AIR utilized the Delphi Method in order to develop estimates of frequency, location, and severity of potential future attacks. AIRWeather is a software product that AIR developed in order to quantify risk of natural disasters. It provides updated weather data for the weather risk management market. AIR also develops risk management software, decision-making software, and other risk-related software that is available for purchase on their website.

2.2.3 Asvaco.
URL: http://www.asvaco.com/

Asvaco provides vulnerability assessment software that creates reports and is compatible with Autodesk, a 2D and 3D software product. Asvaco incorporates all the benefits of the CARVER methodology. CARVER is the acronym of the following set of evaluation criteria: criticality, accessibility, recoverability, vulnerability, effect, and recognizability.

It is used by the Department of Defense, the intelligence community, and the Department of Homeland Security (see reference in Appendix C:5).

2.2.4 Decisioneering. Crystal Ball.
URL: http://www.decisioneering.com/

Although Crystal Ball is not customized as a disaster mitigation program, its features are suitable for performing a probabilistic risk assessment. It is a software package that performs Monte Carlo simulations within the user's spreadsheets. It calculates numerous "what if" cases for individual scenarios and provides a range of possible outcomes and their probability. Decisioneering offers a free 7 day trial version of Crystal Ball on their website.

2.2.5 Defense Threat Reduction Agency (DTRA). Vulnerability Assessment and Protection Option (VAPO).
URL: https://acecenter.cnttr.dtra.mil/acecenter/_login.cfm

Vulnerability Assessment and Protection Option (VAPO) is a software program developed by the Defense Threat Reduction Agency (DTRA) to assess a multi-facility site for threats (i.e., chemical, biological, vehicle ramming, and explosives). It predicts internal and external damage to buildings and equipment along with personal injury and deaths. VAPO provides a geographic information system (GIS) interface for importing and manipulating data and is integrated with HPAC (see reference 3.2.6 on page 56). Structural modeling is quickly done with point-and-click features. For detailed analyses, DTRA offers training courses for this software through the Assessment of Catastrophic Events Center (ACECenter) and provides the software to researchers and government employees without cost.

2.2.6 ESRI.
URL: http://www.esri.com/

ESRI is a private company that provides GIS products, maps, and software for private and public use. In conjunction with FEMA, ESRI has made available quality multi-hazard maps and information that are searchable online by city and type of hazard (http://www.esri.com/hazards/). It has links to FEMA's HAZUS software for hazard analysis and developed an advanced software program, ArcGIS, which is used to create geographic maps. ArcGIS must be installed on a computer in order to use HAZUS (see reference 2.2.7 on page 16) and CATS (see reference 3.2.6 on page 56). A free one month evaluation version of ArcGIS is available from ESRI and FEMA.

2.2.7 Federal Emergency Management Agency. Hazards U.S. (HAZUS).
URL: http://www.fema.gov/plan/prevent/hazus/index.shtm

Hazards U.S. (HAZUS) is a software tool designed to provide individuals, businesses, and communities with information and tools to mitigate natural hazards. HAZUS is a natural hazard loss estimation software program developed by the National Institute of

Building Sciences (NIBS) with funding from FEMA. It allows users to compute estimates of damage and losses from natural hazards using geographical information systems (GIS) technology. Originally designed to address earthquake hazards, HAZUS has been expanded into HAZUS Multi-Hazard (HAZUS-MH), a multi-hazard methodology with new modules for estimating potential losses from wind (including hurricane) and flood hazards. The program is considered to be in a state of development, but a current version of it is available from FEMA. NIBS maintains committees of wind, flood, earthquake, and software experts to provide technical oversight and guidance to HAZUS-MH development. HAZUS-MH uses ArcGIS software to map and display hazard data and the results of damage and economic loss estimates for buildings and infrastructure. The ArcGIS software, developed by ESRI, is required to run HAZUS and is sold separately (see reference 2.2.6 on page 16). Three data input tools have been developed to support data collection: the Inventory Collection Tool, Building Inventory Tool, and Flood Information Tool. The Inventory Collection Tool helps users collect and manage local building data for more refined analyses than are possible with the national level data sets that come with HAZUS. The Building Inventory Tool allows users to import building data from large datasets, such as tax assessor records. The Flood Information Tool helps users manipulate flood data into the format required by the HAZUS flood model. FEMA has also developed a companion software tool called the HAZUS-MH Risk Assessment Tool to produce risk assessment outputs for earthquakes, floods, and hurricanes. The Risk Assessment Tool pulls natural hazard data, inventory data, and loss estimate data into pre-formatted summary tables and text. HAZUS-MH can provide a probability of minor damage, moderate damage, and the destruction of constructed facilities. This information is viewable by occupancy type, type of building, essential facilities, and by user defined facilities. FEMA 433: *Using HAZUS-MH for Risk Assessment* along with the user manuals explains the capabilities of HAZUS-MH.

2.2.8 Federal Emergency Management Agency. Risk Assessment Database. URL: http://www.fema.gov/plan/prevent/rms/rmsp452

FEMA developed a Risk Assessment Database application to support the building assessment process described in FEMA 452 (see reference 3.1.34 on page 42). The Risk Assessment Database is a standalone application that is both a collection tool and a management tool. Assessors can use the tool to assist in the systematic collection, storage, and reporting of assessment data. It has functions, folders, and displays to import and display threat matrices, digital photos, cost data, emergency plans, and certain GIS products. Managers can use the application to store, search, and analyze data collected from multiple assessments. The Risk Assessment Database is initially installed at an organization's headquarters. Referred to as the Manager's Database, it becomes the main access and storage point for future assessment data. When an organization conducts an assessment, a database administrator uses the tool to produce a small temporary database, called the Assessor's Database. References, site plans, GIS portfolios, and other site-specific data that are known about the assessment site are placed into the assessor's database and given to the assessment team. At the end of the assessment, the assessment team combines their data into one database and passes the files back to the database administrator.

2.2.9 Integrated Computer Engineering.
URL: http://www.iceincusa.com/

Integrated Computer Engineering (ICE), a division of American Systems, provides risk management services and software. It has conducted more than 300 risk assessments of Federal, DoD, state and commercial programs using the ISO 9001:2000 industry standard. ICE offers two software products: Risk Radar and Risk Radar Enterprise. Risk Radar is designed for single project risk management and is based on Microsoft Access. Risk Radar Enterprise is designed for enterprise risk management for multiple projects and is a web-based application.

2.2.10 National Institute of Standards and Technology (NIST): Building and Fire Research Laboratory.
URL: http://www.bfrl.nist.gov

The National Institute of Standards and Technology (NIST) is a federal agency in the Department of Commerce that advances productivity through technology and standards. NIST has played a key role in the research and assessment of hurricane damage on the Gulf coast and the investigation of the World Trade Center disaster. Within NIST is the Building and Fire Research Laboratory (BFRL), which studies building materials, construction practices, fire science and safety, and engineering. The laboratory is the source of critical tools –metrics, models, knowledge– used to modernize the building and fire safety communities. Among the many software products developed by BFRL, four are particularly relevant in the assessment of buildings and structures: CONTAM, ALOFT-FT, ASCOS, and FIRST. CONTAM is a software product that assists in the prediction of airflows, contaminant concentration, and personal exposure. It is a multizone indoor air quality and ventilation analysis program that helps determine airflows, contamination dispersal, and exposure prediction. It can also be used to determine the impact of envelope air tightening efforts. ALOFT-FT is a fire plume trajectory model program that predicts the downwind distribution of "smoke particulate and combustion from large outdoor fires" along flat terrain. It was developed for planning the intentional burning of crude oil spills on water. ASCOS analyzes smoke control systems that produce pressure differences in order to limit smoke movement in structural fires. The user provides indoor and outdoor temperatures, the building flow network, and the flows produced by the smoke control system. FIRST predicts fire growth rates using data describing rooms, openings, and the thermo physics of a building. This program can predict fire growth rates and the possible ignition of specified targets. Fire Dynamics Simulator is a "computational fluid dynamics model of fire-driven fluid flow." It solves Navier-Stokes equations with particular focus on smoke and heat transport. Smokeview is a visualization program for the Fire Dynamics Simulator. There are many other programs, both technical and non-technical, provided by NIST.

2.2.11 National Institute of Standards and Technology (NIST): Information Technology Laboratory. Dataplot.
URL: http://www.itl.nist.gov/div898/software/dataplot/

Although Dataplot is not customized for disaster mitigation, its features are suitable for performing a probabilistic risk assessment. It is a free software program used for scientific visualization, statistical analysis, mathematical analysis, and graphical analysis. Several versions are available, each for a different platform: UNIX, VMS, Linux, and Windows. Dataplot is a command-driven system that is written in Fortran-77. This software can be downloaded from the National Institute of Standards and Technology website.

2.2.12 National Institute of Standards and Technology (NIST): Office of Applied Economics. "Building Life-Cycle Cost."
URL: http://www.bfrl.nist.gov/info/software.html

See reference 4.1.3.3 (page 90)

2.2.13 National Institute of Standards and Technology (NIST): Office of Applied Economics. "Cost-Effectiveness Software Tool."
URL: http://www2.bfrl.nist.gov/software/CET/

See reference 4.1.3.4 (page 91)

2.2.14 Palisade. @Risk.
URL: http://www.palisade.com/

Although @Risk is not customized for disaster mitigation, its features are suitable for performing a probabilistic risk assessment. It is a software package that conducts Monte Carlo simulation and sensitivity analysis. @Risk is compatible with Microsoft Excel and provides the probability of possible event outcomes. This software is commonly used to make investment decisions and provides modeling flexibility, standard reports, sensitivity analysis, and error tracking. Palisade offers a 10 day trial version of @Risk on their website.

2.2.15 Risk Management Solutions Incorporated.
URL: http://www.rms.com/

See reference 2.1.22 (page 14)

2.2.16 Sandia National Laboratories. "Risk Assessment Method—Property Analysis and Ranking Tool."
URL: http://www.nwer.sandia.gov/wlp/factsheets/rampart.pdf

Researchers have developed a number of software-based risk assessment tools to model terrorist decision processes as well as risks from natural hazards and other man-made hazards. One such tool is the Risk Assessment Method—Property Analysis and Ranking Tool (RAMPART) software, developed at Sandia National Laboratories with funding from the General Services Administration. RAMPART combines building- and site-specific information with geography-based seismic, weather, and crime data to predict the vulnerability of a building to several categories of consequences due to man-made and natural hazards. In RAMPART, categories of consequences include casualties, damage to property and contents, and loss of use and mission. RAMPART addresses natural hazards (hurricanes, earthquakes, flooding, and winter storms) and several manmade hazards (crime inside the building, crime outside the building, and terrorism). Sandia National Laboratories provides other risk assessment methodology (RAM) programs as well (http://www.sandia.gov/ram/index.htm).

2.2.17 Securac. "Acertus Risk Assessment."
URL: http://www.securac.net/

The Securac risk assessment software program provides information and objectivity in calculating total cost of ownership (TCO), return on investment (ROI), and annual loss expectancy. Following industry standards, it manages risk using qualitative or quantitative assessments. Acertus Risk Assessment can merge internal risk data with national/international databases such as insurance actuarial tables and crime statistics. This software is available for purchase on the Securac website.

2.2.18 Tec-Com Incorporated. RiskWorld.
URL: http://www.riskworld.com/

Tec-Com is a technical communications company that provides an on-line website on the management of environmental, financial, health, and technological risks. The website provides an extensive list of risk related software with web links to the software producers (http://www.riskworld.com/SOFTWARE/SW5SW001.HTM). Most of the software products are produced by private companies and are available for purchase through their websites.

2.3 Hazards Data

2.3.1 Natural Hazards

2.3.1.1 Earthquake Engineering Research Institute and International Association of Earthquake Engineering. World Housing Encyclopedia.
URL: http://www.world-housing.net/

The Earthquake Engineering Research Institute (see reference 3.1.15 on page 37) and the International Association of Earthquake Engineering (see reference 3.1.42 on page 45) have collaborated to produce a web-based encyclopedia of types of housing construction in earthquake prone areas of the world. It provides exceedance of peak ground acceleration of ground movement and a description of the type of housing in the region.

2.3.1.2 EarthQuake Information NETwork.
URL: http://www.eqnet.org/

Funded by FEMA, the Earthquake Information Providers Group (EqIP) disseminates information concerning earthquakes and related topics through the EarthQuake Information NETwork (EQNET). The EQNET website provides links to numerous earthquake resources organized by topics: earthquake information services, structural engineering, geotechnical engineering/engineering geology, seismology/geology/geophysics, disaster management, policy/planning/socioeconomics, agencies and associations, education/professional development, list servers/newsgroups/newsletters/web forums, calendar/conferences, and archives.

2.3.1.3 ESRI.
URL: http://www.esri.com/

See reference 2.2.6 (page 16)

2.3.1.4 Federal Emergency Management Agency.
URL: http://www.fema.gov/

See the description of FEMA in reference 2.1.6 and the *Risk Assessment and Risk Management Guidance Documents* in the references that follow it.

2.3.1.5 High Plains Regional Climate Center and University of Nebraska. "Climate and Weather: Data, Information, and Products Clearinghouse."
URL: http://www.hprcc.unl.edu/clearinghouse/index.html

This clearinghouse provides links to weather data, information, and products. It is divided into three sections: data by time frame, by sector, and by application. Organizations such as the National Climatic Data Center, High Plains Regional Climate Center, and the Department of Agriculture are listed within the links.

2.3.1.6 International Tsunami Information Centre.
URL: http://www.tsunamiwave.info/

The International Tsunami Information Centre (ITIC) was established by the United Nations Intergovernmental Oceanographic Commission (IOC) in order to mitigate tsunami hazards. The ITIC website provides information about tsunamis and data concerning past tsunamis.

2.3.1.7 Michigan Technological University Volcanoes Page.
URL: http://www.geo.mtu.edu/volcanoes/

The Michigan Technological University Volcanoes Page provides information, links, and data about volcanic activity, mitigation, maps, and other interests. This site is useful for finding sites related to volcanoes such as the United States Geological Survey, NASA, the Smithsonian Institute, and journals related to volcanology.

2.3.1.8 Mid-America Earthquake Center.
URL: http://mae.ce.uiuc.edu/

The Mid-America Earthquake Center (MAE) was established by the National Science Foundation (NSF) and provides education, research, and outreach concerning earthquake hazards. Their website has a list of publications associated with the MAE Center and provides several software products developed by the MAE Center for evaluating earthquake hazards. The publications listed discuss engineering research and tend to be technical in nature. Unfortunately, they are not available to download directly from the website.

2.3.1.9 NASA. "Disaster! Finder."
URL: http://disasterfinder.gsfc.nasa.gov/

The NASA "Disaster! Finder" website provides a list of 6 topics to browse: disaster management (e.g., mitigation, preparedness, and relief), disciplines (e.g., climate, GIS, and insurance), general (e.g., studies, workshops, and collections), organizations (e.g., public and private), systems (e.g., GIS software, emergency management systems, and observing systems), and type of disaster (e.g., avalanche, drought, earthquake, and flood). Each topic contains a list of subtopics with descriptions and links to related resources.

2.3.1.10 National Oceanic and Atmospheric Administration (NOAA).
URL: http://www.noaa.gov/

The NOAA is part of the Department of Commerce and is responsible for predicting changes in and informing the public about the earth's environment as it relates to coastal and marine resources. The National Weather Service; National Environmental Satellite, Data, and Information Service (NESDIS); the National Ocean Service; and the Marine and Aviation Operations are part of the NOAA. Current and historical weather, ocean, and climate information (e.g., precipitation amount, storms, wind, and pressure) is available from the NOAA website and can be viewed in daily, monthly, or annual records. While some information requires a fee, many of these products are free to the public. The NOAA provides information and data on tsunamis (http://nctr.pmel.noaa.gov/ and http://wcatwc.arh.noaa.gov/ and http://www.prh.noaa.gov/ptwc/), hurricanes (http://hurricanes.noaa.gov/ and http://www.nhc.noaa.gov/), coastal data (http://www.csc.noaa.gov/), severe thunderstorms (http://www.nssl.noaa.gov/), climate and weather (http://www.weather.gov/om/), floods (http://www.nws.noaa.gov/oh/hic/archive/), and fire weather (http://www.spc.noaa.gov/fire/). Another site of the NOAA that provides data and imagery on natural hazards is located at http://www.ngdc.noaa.gov/seg/hazard/hazards.shtml.

2.3.1.11 National Oceanic and Atmospheric Administration. Storm Data (SD).
URL: http://www5.ncdc.noaa.gov/pubs/publications.html

See reference 3.3.2.13 (page 67)

2.3.1.12 National Sea Grant Network. "HazNet."
URL: http://www.haznet.org/

HazNet is the National Sea Grant Network web site and is dedicated to reducing loss of life and property due to coastal hazards. It provides information and research about hazard mitigation, research projects, outreach, and other state Sea Grant institutions to contact or visit concerning coastal hazards. It provides abstracts of research projects, but the project report is not available directly from the website.

2.3.1.13 National Weather Service.
URL: http://www.nws.noaa.gov/

Being the primary information source for weather in the U.S., The National Weather Service (NWS) provides information that forecasters use to make weather predictions. The NWS provides weather, hydrologic, and climate forecasts, not only for the U.S., but also for surrounding areas. River levels, safety, warnings, radar information, satellite data, and an air quality assessment that provides ozone concentration levels are easily searchable by map and location name. Covering 1995 through 2004, statistics of natural disasters are grouped by number of fatalities, injuries, amount of property damage, and crop damage.

2.3.1.14 National Weather Service: Office of Climate, Water, and Weather Services.
URL: http://www.nws.noaa.gov/om/hazstats.shtml

The Office of Climate, Water, and Weather Services provides a comprehensive list of natural disaster statistics. The information is available for state and national summaries as well as by the type of natural hazard. Currently, the data is available for the years of 1995 through 2004. The summary tables provide the number of fatalities and injuries along with the amount of property damage and crop damage. The data by natural hazard type provides the number of fatalities by state and type of activity/location.

2.3.1.15 National Oceanic and Atmospheric Administration (NOAA) Satellite and Information Service: National Environmental Satellite, Data, and Information Service (NESDIS)
URL: http://www.nesdis.noaa.gov/

Providing global environmental data from satellites and other sources, the National Environmental Satellite, Data, and Information Services (NESDIS) is made up of several organizations: the National Climatic Data Center, the National Geophysical Data Center, the National Oceanographic Data Center, the National Coastal Data Development Center, and the Office of Satellite Operations. The National Climatic Data Center (NCDC) contains the world's largest archive of climate data and operates the World Data Center for Meteorology, which is co-located at NCDC. Envisioned to be the most comprehensive and accessible source of weather data, the NCDC distributes climate, satellite, and radar data, some of which is free to the public. The National Oceanographic Data Center (NODC) manages and preserves long-term and short-term oceanographic data. There are several ocean-data web applications that can be accessed through NODC: Acoustic Doppler Current Profiler Data, Argo Profiling Float Data, Coastal Buoy data, Coastal Water Temperature Guide, Global Temperature-Salinity Profile Program, Joint Archive for Sea Level, Ocean Time Series Data Base, and Satellite Data from NODC.

2.3.1.16 Natural Environment Research Council. "Tsunami Risks Project."
URL: http://www.nerc-bas.ac.uk/tsunami-risks/

This site is provided by the Natural Environment Research Council of Coventry University and University College London. It has brief sections on the cause, physics, consequence, risk, and mitigation of tsunamis events. It also provides discussions about past tsunami events and their effects.

2.3.1.17 Pacific Disaster Center.
URL: http://www.pdc.org

The Pacific Disaster Center (PDC) provides policy support and information products concerning disasters in the Asia Pacific region. Their webpage provides disaster maps (http://www.pdc.org/atlas) and a list of PDC publications, which are not accessible through their website.

2.3.1.18 Swiss Re. Twister! The Professional Reinsurer's Perspective.
URL: http://www.swissre.com/

This publication provides occurrence and frequency information on tornados. It focuses primarily on the United States and discusses the impact of tornados. It proposes that a greater understanding of tornados will help prevent deaths, injuries, and damage. It briefly describes the Fujita wind damage scale, which is commonly used to measure the severity of a tornado.

2.3.1.19 Tropical Storm Risk Consortium.
URL: http://tsr.mssl.ucl.ac.uk/

The Tropical Storm Risk (TSR) Consortium was developed from a project of the United Kingdom and draws information from numerous entities of expertise. Their website contains viewable publications and information about present and past tropical storms occurring in various parts of the world.

2.3.1.20 United States Geological Survey.
URL: http://www.usgs.gov/

The United States Geological Survey (USGS) is a federal agency in the Department of the Interior that manages and provides information concerning water, biological, energy, and mineral resources. It has produced background information for the nation concerning seven types of natural hazards: floods, earthquakes, landslides, volcanic eruptions, coastal storms and tsunamis, wildfires, and outbreaks of disease in wildlife populations (http://www.usgs.gov/themes/hazard.html). Researchers at the USGS Coastal and Marine Geology Program are also developing models to predict the occurrences, severity, and consequences of natural disasters (http://pubs.usgs.gov/fs/natural-disasters/index.html). Live updates of natural disasters occurring around the world can be viewed at http://nhss.cr.usgs.gov/.

2.3.1.21 United States Geological Survey: Earthquakes Hazard Program.
URL: http://earthquake.usgs.gov/

The USGS National Earthquake Information Center (NEIC: http://wwwneic.cr.usgs.gov/), National Seismic Hazard Mapping Project (http://earthquake.usgs.gov/hazmaps), and Advanced National Seismic System (ANSS: http://earthquakes.usgs.gov/anss) provide earthquake data and hazard maps through the USGS Earthquakes Hazard Program. The national and global earthquake hazard data are available by zip code or by latitude and longitude. The Earthquakes Hazard Program provides a site dedicated to the preparedness and response to seismic events.

2.3.1.22 United States Geological Survey: Marine and Coastal Geology Program.
URL: http://marine.usgs.gov/

The USGS Coastal and Marine Geology Program balances coastal and marine resources. It has several projects associated with hurricane and coastal storm prediction. The Coastal Classification Mapping Project (http://coastal.er.usgs.gov/coastal-classification/index.html) characterizes and classifies pre-storm ground conditions for states located along the Gulf of Mexico that, when combined with data about beach stability and prior storm impact studies (http://coastal.er.usgs.gov/hurricanes/index.html), provide indications of an area's vulnerability to hurricanes or other extreme coastal storms. Information on this site is categorized by topic, region, and content.

2.3.1.23 United States Geological Survey: National Geospatial Program Office.
URL: http://www.usgs.gov/ngpo/

The National Geospatial Program Office brings the National Map, Geospatial One-Stop, and the Federal Geographic Data Committee into one program office. The National Map (http://nationalmap.gov/) provides a common set of base information that describes the earth's surface and locates geographic features. There are multiple layers to this map that include both political and geographical information. The Geospatial One-Stop (http://www.geodata.gov) has links to information and has the ability to search multiple databases. It has fire mapping, administrative boundaries, facility information, and agricultural information as well as atmospheric, economic, demographic, elevation, transportation, utility, and water data. The Federal Geographic Data Committee (http://www.fgdc.gov/) promotes the coordinated development, use, and dissemination of geospatial data. Their website provides the National Spatial Data Infrastructure, which searches numerous global databases.

2.3.1.24 United States Geological Survey: Tsunami and Earthquake Research at the USGS.
URL: http://walrus.wr.usgs.gov/tsunami/

This USGS website provides information and reports from current research being conducted on tsunamis and earthquakes. It provides models, animations, and analyses of past tsunamis and earthquakes. Other links that provide further information are listed on their website.

2.3.1.25 United States Geological Survey: USGS Landslide Hazards Program.
URL: http://landslides.usgs.gov/

The USGS Landslide Hazards Program is dedicated to reducing long-term losses from landslide hazards by advancing knowledge and mitigation. It conducts research, produces reports, and responds to emergencies and disasters. The "National Landslide Hazards Mitigation Strategy- A Framework for Loss Reduction" is one publication that can be found on their site (http://pubs.usgs.gov/circ/c1244/c1244.pdf). It discusses the probability and damage caused by a landslide and provides a comprehensive community/government strategy for reducing losses. There are a number of other useful publications on landslides through their website.

2.3.1.26 United States Geological Survey: Volcano Hazards Program.
URL: http://volcanoes.usgs.gov/

The USGS Volcano Hazards Program provides information on reducing volcanic risk, volcano monitoring, emergency planning, warning schemes, and disaster assistance. It also has a list of resources for facts, photos, videos, and reports. Another USGS site related to volcanoes is at http://www.usgs.gov/themes/volcano.html.

2.3.1.27 University of Colorado-Boulder: Center for Science and Technology. "Extreme Weather Sourcebook."
URL: http://sciencepolicy.colorado.edu/sourcebook/

This website contains a sourcebook of economic and societal impacts caused by hurricanes, floods, tornadoes, lightning, and other selected weather. The damage is listed by state and provides a ranking and dollar amount for the damage caused by each type of weather condition.

2.3.1.28 World Health Organization, Collaborating Centre for Research on the Epidemiology of Disasters. EM-DAT Emergency Disasters Data Base.
URL: http://www.em-dat.net/

This disaster database was created with the support of the World Health Organization (http://www.cred.be/cred1/index.htm) and the Belgian Government in order to evaluate disaster vulnerability and disaster preparedness. The data is downloadable and searchable by region, country, period, and type/group of disaster. The database contains over 12 800 mass disasters in the world from 1900 to the present, including both natural and manmade disasters.

2.3.2 Man-Made Hazards

2.3.2.1 Chemical Safety Board (CSB).
URL: http://www.csb.gov/

The U.S. Chemical Safety Board (CSB) is a federal agency that is responsible for investigating industrial chemical accidents. It does not issue fines, but makes recommendations to public and private entities. The CSB website contains reports on current and completed investigations, safety publications, and recommendations by the CSB.

2.3.2.2 Environmental Protection Agency (EPA).
URL: http://www.epa.gov/

The Environmental Protection Agency (EPA) protects the health and environment of the nation. It conducts research, publishes articles, and develops and enforces regulations. Their website has facts and publications on acid rain, air quality, asbestos, water quality, climate change, hazardous waste, mercury levels, mold, oil spills, ozone, pesticides, and radon.

2.3.2.3 Environmental Protection Agency (EPA). "Toxics Release Inventory (TRI) Program."
URL: http://www.epa.gov/tri/

The Toxics Release Inventory (TRI) Program is a database of toxic chemical releases and other waste management activities reported annually to the Environmental Protection Agency (EPA). The data includes location, type of release, and type of chemical.

2.3.2.4 ESRI.
URL: http://www.esri.com/

See reference 2.2.6 (page 16)

2.3.2.5   Federal Bureau of Investigation. Annual Uniform Crime Report.
URL: http://www.fbi.gov/publications.htm

The Federal Bureau of Investigation (FBI) defends against terrorist, foreign intelligence, and criminal acts in the United States. The FBI website contains information on cyber crimes, public corruption, organized crime, and environmental crimes. Crime statistics are provided through the annual *Uniform Crime Report*. This report details crimes in the U.S. by region, weapon, and incidence. This extensive report provides current and archive crime data.

2.3.2.6   Federal Bureau of Investigation. Terrorism in the United States.
URL: http://www.fbi.gov/publications.htm

The FBI publishes a periodic report that provides descriptions and selected statistics of terrorism incidences in the United States. It includes information about the perpetrators of terrorist incidents, the number of terrorist acts committed, and the number of acts in each region of the U.S.

2.3.2.7   Federal Emergency Management Agency.
URL: http://www.fema.gov/

See reference 2.1.6 (page 9)

2.3.2.8   Marlatt, Greta E. Chemical, Biological, and Nuclear Terrorism/Warfare: A Bibliography. Dudley Knox Library: Naval Postgraduate School, September 2003.
URL: http://www.nps.edu/Library/Research/Bibliographies/CBNTerror/CBNTerrorBib.pdf.pdf

This text is a bibliography on terrorism and warfare. It has three primary sections: chemical, biological, and nuclear terrorism/warfare. Each of these sections has four subsections: periodicals, books, technical reports, and websites.

2.3.2.9   National Fire Protection Association.
URL: http://www.nfpa.org

The National Fire Protection Association (NFPA) is a private organization that is committed to reducing life-threatening hazards through research and standardized codes. The NFPA provides safety, statistical, and fire equipment information to the public; nearly every building is affected by NFPA standards and documents. The different kinds of statistical information include data on trends, causes, history, and services as it relates to fires in the United States. The NFPA provides codes and standards in publication form and on the Internet; access to the comprehensive fire code publication requires membership. Publications include the National Fire Alarm Code, Designer's Guide to Automatic Sprinkler Systems, and NFPA Catalog, which contains fire, electrical, and building safety products. The NFPA provides fact sheets that are categorized into Home

Safety, Safety in other occupancies, Homeland Security, Fire protection equipment, seasonal safety, Vehicle fires/gas and fuel safety, Oil refineries, and Wildland fires. Each category is then subcategorized into specific issues. For example, under Vehicle fires/gas and fuel safety is Gasoline at home, Propane safety, and Service station safety. Although the fact sheets do not provide the raw data for analysis, they contain facts and figures.

2.3.2.10 National Memorial Institute for the Prevention of Terrorism (MIPT)
   URL: http://www.mipt.org/

See reference 3.3.2.11 (page 67)

2.3.2.11 National Memorial Institute for the Prevention of Terrorism. "Terrorism Knowledge Base."
   URL: http://www.tkb.org/Home.jsp

The National Memorial Institute for the Prevention of Terrorism (MIPT) and the RAND Corporation coordinated the Terrorism Knowledge Base. The project has two databases: the RAND Terrorism Chronology Database and the RAND-MIPT Terrorism Incident Database. The former records international terrorist incidents that occurred between 1968 and 1997; the later records domestic and international terrorist incidents occurring from 1998 to the present. The website includes a great deal of information about terrorist groups, incidents, cases, leaders, their countries, and other issues. See reference 2.1.21 on page 13 and 3.3.2.11 on page 67.

2.3.2.12 Swiss Re.
   URL: http://www.swissre.com/

See reference 3.3.2.22 (page 69)

2.3.2.13 United States Army Training and Doctrine Command, Deputy Chief of Staff for Intelligence, Assistant Deputy Chief of Staff for Intelligence, Threats. A Military Guide to Terrorism in the Twenty-First Century. U.S. Army DCSINT Handbook No. 1. Fort Leavenworth, KS: August 15, 2005.
   URL: http://fas.org/irp/threat/terrorism/

This publication details terrorism threats to the U.S. military, but can be useful to any organization that might have contact with a terrorist group. It has descriptions of improvised explosive devices, weapons of mass destruction, terrorist firearms, terrorist tactics, and contains a history of terrorism. This publication is useful for familiarization of terrorist practices and activities. DCSINT Handbook No. 1.02: *Cyber Operations and Cyber Terrorism* is a supplement to this Handbook (see reference 2.3.2.14 on page 30).

2.3.2.14 United States Army Training and Doctrine Command, Deputy Chief of Staff for Intelligence, Assistant Deputy Chief of Staff for Intelligence, Threats. Cyber Operations and Cyber Terrorism. U.S. Army DCSINT Handbook No. 1.02. Fort Leavenworth, KS: August 15, 2005.
URL: http://www.fas.org/irp/threat/terrorism/sup2.pdf

DCSINT Handbook No. 1.02: *Cyber Operations and Cyber Terrorism* is a supplement to Handbook No. 1 (see reference 2.3.2.13 on page 29). It discusses cyber terrorism, its threat to U.S. infrastructure, and its threat to the military. The objective of cyber terrorism includes the loss of availability, loss of confidentially, and physical destruction.

2.3.2.15 United States Census Bureau.
URL: http://www.census.gov

The Census Bureau is part of the Department of Commerce and is a leading source of information concerning the U.S. economy and people. This agency provides downloadable data in tables or viewable on maps. Some of the information includes: personal income, population, poverty, insurance, housing, employment, economic indicators, and more. The data is viewable for the nation or for specific cities, counties, states, or regions. The Census Bureau is a primary source of statistical data on the nation.

2.3.2.16 United States Census Bureau. Statistical Abstract of the United States.
URL: http://www.census.gov/statab/www/

The *Statistical Abstract of the United States* contains numerous social and economic statistics for the United States, which include population, law enforcement, education, health, prison, court, and construction statistics. This publication is also a guide to sources of data from other federal and private agencies.

2.3.2.17 United States Department of Justice: Bureau of Justice Statistics.
URL: http://www.ojp.usdoj.gov/bjs/

The Bureau of Justice Statistics provides statistics on crime and victims, criminal offenders, law enforcement, prosecution, courts and sentencing, and other topics. It contains both state and national level statistics. The Bureau of Justice Statistics lists additional sources of information for each subject.

2.3.2.18 United States Department of Transportation: Bureau of Transportation Statistics.
URL: http://www.bts.gov/

The Bureau of Transportation Statistics (BTS) provides data on transportation. The data is searchable by mode of transportation (i.e., aviation, maritime, highway, transit, rail, pipeline, or bike/pedestrian) or by subject (i.e., safety, freight transport, passenger travel, infrastructure, economic/financial, social/demographic, energy, environment, and national security). Various other statistics are available through the BTS TransStats websites (http://transtats.bts.gov/). Also see the U.S. Department of Transportation in reference 3.3.1.34.

2.3.2.19 United States Fire Administration.
URL: http://www.usfa.dhs.gov/

The United States Fire Administration (USFA) is part of the Department of Homeland Security that provides education, research, and data associated with economic and life losses due to fires. Their website contains fire statistics, the National Fire Incident Reporting System (NFIRS), and the National Fire Department Census Database. Fire statistics on the USFA website includes data on arson, wildfire, seasonal fires, fire departments, and firefighters. The NFIRS is a standard national reporting system for fires and other incidents that U.S. fire departments respond to. The National Fire Department Census Database is an online address listing of U.S. fire departments.

3 Risk Management Resources

The risk management step focuses on identification of risk mitigation strategies. This step uses information from the risk assessment (e.g., estimates of the value of the facility's assets and the consequences of an event occurring) to identify engineering, management and financial strategies to mitigate those consequences. The costs of implementing the alternative risk mitigation strategies and the associated reductions in consequences are also produced as part of this step.

Section 3.1 covers risk management guidance documents. The emphasis in Section 3.1 is on establishing a technical foundation for formulating combinations of risk mitigation strategies that address the hazards identified in the risk assessment and identifying key resource documents that provide guidance on how to formulate multi-hazard risk mitigation strategies. Each strategy is composed of multiple approaches for addressing hazards identified in the risk assessment. These approaches focus on hazard mitigation for a specific system or collection of systems and components, as well as facility and site-related elements. Documents abstracted include the FEMA *Risk Management Series* of publications, as well as publications from the AIA, ASCE, CII, GAO, GSA, IBHS, NFPA, NRC, and the NSTC.

Section 3.2 focuses on risk management software. Coverage of software tools focuses on their value in compiling, organizing, and evaluating information and data needed to formulate alternative combinations of risk mitigation strategies.

Section 3.3 covers guidance documents for cost-related issues; it is divided into two subsections: mitigation costs and event-related losses. Documents abstracted are designed to help the reader compile information on the amount and timing of investment costs, operating costs, maintenance and repair costs, and consequences for each alternative combination of risk mitigation strategies.

Section 3.4 covers software tools that are useful in developing estimates of mitigation costs and event-related losses.

3.1 Risk Management Guidance Documents

3.1.1 Advanced Materials and Processes Technology Information Analysis Center. URL: http://ammtiac.alionscience.com/

The Advanced Materials and Processes Technology Information Analysis Center (AMMTIAC) is chartered by the Department of Defense to provide data and information. Their website provides a list of their publications, which are available for purchase. AMPTIAC Quarterly is a publication that provides research on advanced materials and processes; many of the articles discuss the protection of buildings and structures. AMMTIAC also provides an Inquiry Service, which provides free advice and technical solutions via telephone or email.

3.1.2 American Institute of Architects. Security Planning and Design: A Guide for Architects and Building Design Professionals. Hoboken, NJ: Wiley & Sons, 2003.

This text provides current information on security planning in new and existing facilities. It discusses security design concepts, security evaluation, planning, building hardening, security technology, biochemical and radiological protection, and putting security into practice.

3.1.3 American Management Association. Facility Manager's Emergency Preparedness Handbook.
URL: http://www.amanet.org/

In 2003, the American Management Association (AMA) published the *Facility Manager's Emergency Preparedness Handbook*. This handbook is intended as a reference for emergency preparedness planning. It provides guidelines, tools, and checklists to facility managers to prepare for several types of emergencies: terrorism, fire emergency, lockout, and workplace violence.

3.1.4 American Red Cross.
URL: http://www.redcross.org/

The American Red Cross is a private emergency response organization, financed primarily through donations. The Red Cross responds to natural disasters, hazardous spills, transportation accidents, explosions, and any other natural or man-made disaster. Their website is useful for preparation in responding to a disaster or emergency, and it provides information on all their services. It discusses topics such as water treatment, food safety, generators, financial losses, chemical emergencies, earthquakes, fires, floods, heat waves, hurricanes, mudslides, storms, tornadoes, volcanoes, and fires. This organization targets private individuals, who have been or will be affected by a disaster. The Red Cross federated with the Red Crescent Society has developed the ProVention Consortium, which aims at reducing global risk of disaster:
http://www.proventionconsortium.org/.

3.1.5 American Society for Industrial Security (ASIS).
URL: http://www.asisonline.org/

American Society for Industrial Security (ASIS) develops educational programs and materials that address security issues. ASIS administers three certification programs: the Certified Protection Professional (CPP), the Physical Security Professional (PSP), and the Professional Certified Investigator (PCI). It also hosts the ASIS International Annual Seminar and Exhibits.

3.1.6   American Society of Civil Engineers.
         URL: http://www.asce.org/asce.cfm

Founded in 1852, the American Society of Civil Engineers (ASCE) promotes engineering to build a better quality of life. It has numerous publications, which include a monthly magazine, *Civil Engineering*; a monthly newspaper, *ASCE News*; the quarterly *Geo-Strata*; numerous journals; and a variety of books. It has produced a number of texts on infrastructure vulnerability. ASCE maintains a codes and standards program that is divided into six specialized committee groups: Architectural Engineering Institute, Construction Institute, Environmental and Water Resources Institute, Geo-Institute, Structural Engineering Institute, and the Transportation and Development Institute.

3.1.7   American Society of Civil Engineers. Flood Resistant Design and Construction. SEI/ASCE 24-05. Reston, VA: American Society of Civil Engineers, 2006.

This standard establishes minimum requirements for flood-resistant design and construction of new facilities in flood zones. It addresses design, materials, dry and wet flood proofing, utilities, means of egress, and accessory structures in areas subject to flooding. This document is available for purchase from the American Society of Civil Engineers.

3.1.8   American Society of Civil Engineers. Minimum Design Loads for Building and Other Structures. SEI/ASCE 7-05. Reston, VA: American Society of Civil Engineers, 2006.

This text provides the latest "consensus requirements for soil, flood, wind, snow, rain, ice, and earthquake loads, and their combinations, that are suitable for inclusion in building codes and other documents." This text is referenced in the International Building Code (see reference 3.1.43 on page 45) and the 2002 NFPA 5000 Building and Construction Safety Code.

3.1.9   American Society of Civil Engineers. Prestandard and Commentary for the Seismic Rehabilitation of Buildings. FEMA 356. Washington DC: Federal Emergency Management Agency, November 2000.
         URL: http://www.nehrp.gov/info/PDF/FEMA356/FEMA356.pdf

This prestandard provides preliminary provisions for the rehabilitation of buildings to improve seismic performance. The intended audience includes code officials, building owners, and design professionals. It discusses building configuration, component properties, primary and secondary elements, structural strengthening, design requirements, and many other topics. This publication is technical in nature.

3.1.10 American Society of Civil Engineers. Structural Design for Physical Security. Reston, VA: American Society of Civil Engineers.

This text is a guide to protect civil structures from the effects of explosive blasts. It is written primarily for structural engineers and discusses "methods by which structural loadings are derived for the determined threat, the behavior and selection of structural systems, the design of structural components, the design of security doors, the design of utility openings, and the retrofitting of existing structures."

3.1.11 Blue Ribbon Panel on Bridge and Tunnel Security. Recommendations for Bridge and Tunnel Security. Federal Highway Administration, September 2003.
URL: http://www.tisp.org/files/pdf/aashto_fhwa_bdge_tunnel_security.pdf

The Federal Highway Administration (FHWA) is responsible for improving highway performance and provides "leadership in defining future transportation" (http://www.fhwa.dot.gov/). A Blue Ribbon Panel of bridge and tunnel experts was convened at the FHWA to produce *Recommendations for Bridge and Tunnel Security*. The panel was jointly sponsored by the FHWA and the American Association of State Highway and Transportation Officials (AASHTO). This publication is intended to recommend policies and actions to reduce the probability of catastrophic structural damage to bridges and tunnels.

3.1.12 Building Seismic Safety Council. NEHRP Recommended Provisions for Seismic Regulations for New Buildings and other Structures. FEMA 450. Washington DC: Federal Emergency Management Agency, 2003.
URL: http://www.nehrp.gov/info/index.html

This publication presents criteria for the design and construction of structures to resist ground motion caused by earthquakes. The intent of these provisions is to create an environment where there is a low likelihood of structural collapse. Buildings designed by these provisions are likely to suffer repairable damage; however, the cost of repairs may not be economical. The provisions apply to the design and alteration of new and existing structures. There are two parts and two additional sections to FEMA 450: Provisions, Commentary, MCE Maps, and Long Period Maps.

3.1.13 Cauffman, Stephen A. and H. S. Lew. Standards of Seismic Safety for Existing Federally Owned and Leased Buildings. NISTIR 6762. ICSSC RP 6. Gaithersburg, MD: National Institute of Standards and Technology, January 2002.
URL: http://www.nehrp.gov/info/PDF/NISTIR_6762.pdf

This publication provides federal agencies with minimum and extended standards for the evaluation and mitigation of seismic events on buildings. In this document, Life-Safety is the minimum acceptable performance, and Immediate Occupancy is considered to be an extended level of performance. These definitions are founded in FEMA 310 (FEMA 310 has been replaced by ASCE 31-03 in reference 2.1.3) and FEMA 356 (see reference 3.1.9 on page 35). Topics in this document include rehabilitation, component modification, irregularities, mitigation methods, and historic buildings.

3.1.14 Construction Industry Institute. Implementing Project Security Practices. Implementation Resource BMM-3. July 2005.

This guide was developed by the Construction Industry Institute (CII) in order to determine best practices for project security of industrial projects. It provides assistance in transitioning from security research to implementation and is most appropriate for those who are managing security issues from the very initial phases of building design to the final stages of construction. (See reference 3.1.63 on page 50)

3.1.15 Earthquake Engineering Research Institute.
URL: http://www.eeri.org/

This organization is a group of technical and social scientists committed to reducing earthquake risk by improving behavior and designs. It has several publications and a professional journal titled *Earthquake Spectra* that are available for purchase.

3.1.16 Earthquake Engineering Research Institute. Action Plan for Performance Based Seismic Design. FEMA 349. Washington DC: Federal Emergency Management Agency, April 2000.
URL: http://www.nehrp.gov/information.html

This publication provides a plan to implement performance based seismic design, which is a building design that has "predictable and reliable performance in earthquakes." This method uses a "sliding scale" that allows a building to be designed to perform at certain level in an earthquake of a specified severity. Stakeholders can select a level of performance and more easily quantify expected risks to their buildings.

3.1.17 Eidinger, John M., and Ernesto A Avila. Guidelines for the Seismic Evaluation and Upgrade of Water Transmission facilities. Reston, VA: American Society of Civil Engineers, 1999.

This text addresses the vulnerability of water transmission facilities in the case of a seismic event. It provides guidelines for "evaluation and retrofit of water transmission infrastructure including aqueducts, tunnels, canals, buried pipelines, elevated pipelines and their appurtenances." It discusses current practices and past performance.

3.1.18 Federal Alliance for Safe Homes (FLASH).
URL: http://www.flash.org/

FLASH is a non-profit organization that promotes disaster safety and mitigation for residential homes. It provides home safety information that discusses preparing and responding to earthquakes, extreme temperatures, floods, hail, hurricanes, lightning, power outage, rip currents, terrorism, thunderstorms, tornadoes, tsunamis, and wildfire. FLASH also provides online animations of securing and reinforcing home roofs, walls, windows, doors, foundation, and safe landscaping.

3.1.19 Federal Emergency Management Agency (FEMA).
URL: http://www.fema.gov/

See reference 2.1.6 (page 9) and the FEMA references that follow it

3.1.20 Federal Emergency Management Agency. Building Design for Homeland Security. FEMA E 155. Washington DC: Federal Emergency Management Agency.
URL: http://www.fema.gov/plan/prevent/rms/index.shtm

FEMA has created a course, E 155: *Building Design for Homeland Security*, which draws on FEMA 426 and FEMA 452. The course familiarizes students with assessment methodologies available to identify the relative level of risk for a variety of threats, including blast, chemical, biological, and radiological weapons. This course is designed for engineers, architects, and building officials.

3.1.21 Federal Emergency Management Agency. Communicating with Owners and Managers of New Buildings on Earthquake Risk: A Primer for Design Professionals. FEMA 389. Washington, DC: Federal Emergency Management Agency, January 2004.
URL: http://www.fema.gov/plan/prevent/rms/index.shtm

FEMA 389: *Communicating with Owners and Managers of New Buildings on Earthquake Risk: A Primer for Design Professionals* is a FEMA Risk Management Series publication. It was developed to aid in educating building owners and managers about seismic risk management tools. The guide discusses site location, seismic hazards, surface fault rupture, soil hazards, and other hazards posed by seismic events. This detailed document provides illustrations, data, checklists, a risk management plan, case studies, and specifications.

3.1.22 Federal Emergency Management Agency. Design Guidance for Multi-Hazard Shelters. FEMA 453. Washington, DC: Federal Emergency Management Agency.
URL: http://www.fema.gov/plan/prevent/rms/index.shtm

FEMA 453 of the FEMA Risk Management Series publications provides design information on shelters and safe rooms for constructed facilities. It is in support of the All-Hazards preparedness objective established by Homeland Security Presidential Directive/HSPD-8 (see reference in Appendix C:3 on page 123). The information in FEMA 453 is applicable to the work place, home, or community building. These shelters are temporary safe rooms that are not intended to provide continuous protection and are effective only if there is warning of an approaching disaster.

3.1.23 Federal Emergency Management Agency. Design Guide for School Safety against Earthquakes, Floods, and High Winds. FEMA 424. Washington, DC: Federal Emergency Management Agency, January 2004.
URL: http://www.fema.gov/plan/prevent/rms/index.shtm

FEMA 424: *Design Guide for School Safety against Earthquakes, Floods, and High Winds,* is a manual of the FEMA Risk Management Series publications designed to make available guidance for the protection of school buildings from natural disasters. It primarily addresses grade schools and considers the threat of earthquakes, floods, and high winds. The manual emphasizes two methodologies: multi-hazard design and performance-based design. Discussing the safety of occupants along with economic and social losses, the guide considers past and present school designs and school needs.

3.1.24 Federal Emergency Management Agency. FEMA Reports Prepared for SAC Steel Project. FEMA 350 Series. Washington DC: FEMA, 2000.
URL: http://www.nehrp.gov/information.html

This series of reports includes 10 publications that detail structural design of buildings to withstand earthquakes. It includes the following documents:

- FEMA 350: Recommended Seismic Design Criteria for New Steel Moment-Frame Buildings
- FEMA 351: Recommended Seismic Evaluation and Upgrade Criteria for Existing Welded Steel Moment-Frame Buildings
- FEMA 352: Recommended Post earthquake Evaluation and Repair Criteria for Welded Steel Moment-Frame Construction for Seismic Applications
- FEMA 354: A Policy Guide to Steel Moment-Frame Construction
- FEMA 355A: State of the Art Report on Base Materials and Fracture
- FEMA 355B: State of the Art Report on Welding and Inspection
- FEMA 355C: State of the Art Report on Systems Performance of Steel Moment Frames Subject to Earthquake Ground Shaking
- FEMA 355D: State of the Art Report on Connection Performance
- FEMA 355E: State of the Art Report on Past Performance of Steel Moment-Frame Buildings in Earthquakes
- FEMA 355F: State of the Art Report on Performance Prediction and Evaluation of Steel Moment-Frame Structures

3.1.25 Federal Emergency Management Agency. Incremental Seismic Rehabilitation of Hospital Buildings. FEMA 396. Washington, DC: Federal Emergency Management Agency, December 2003.
URL: http://www.fema.gov/plan/prevent/rms/index.shtm

FEMA 396: *Incremental Seismic Rehabilitation of Hospital Buildings* of the FEMA Risk Management Series publications is a manual that provides information to assess the seismic vulnerability of healthcare facilities. The manual contains three parts: Part A,

Critical Decisions for Earthquake Safety in Hospitals; Part B, Planning and Managing the Process for Earthquake Risk Reduction in Existing Hospital Buildings; and Part C, Tools for Implementing Incremental Seismic Rehabilitation in Existing Hospital Buildings. Each part is directed toward different administrative functions of a healthcare facility.

3.1.26 Federal Emergency Management Agency. Incremental Seismic Rehabilitation of Multifamily Apartment Buildings. FEMA 398. Washington, DC: Federal Emergency Management Agency, February 2004.
URL: http://www.fema.gov/plan/prevent/rms/index.shtm

FEMA 398: *Incremental Seismic Rehabilitation of Multifamily Apartment Buildings* is a manual that provides owners of multifamily buildings information to implement a seismic rehabilitation program. The manual consists of three parts: Part A, Critical Decisions for Earthquake Safety in Multifamily Buildings; Part B, Planning and Managing the Process for Earthquake Risk Reduction in Existing Multifamily Buildings; and Part C, Tools for Implementing Incremental Seismic Rehabilitation in Existing Multifamily Buildings. Each part has a different audience: senior executives, board members, policy makers, facility managers, risk managers, financial managers and those who are responsible for the implementation of a seismic program.

3.1.27 Federal Emergency Management Agency. Incremental Seismic Rehabilitation of Office Buildings, FEMA 397. Washington, DC: Federal Emergency Management Agency, December 2003.
URL: http://www.fema.gov/plan/prevent/rms/index.shtm

FEMA 397: *Incremental Seismic Rehabilitation of Office Buildings* is a manual that provides owners of office buildings with information to evaluate the vulnerability of their buildings. This manual contains three sections: Part A, Critical Decisions for Earthquake Safety in Office buildings; Part B, Planning and Managing the Process for Earthquake Risk Reduction in Existing Office Buildings; and Part C, Tools for Implementing Incremental Seismic Rehabilitation in Existing Office Buildings. Each part has a specific audience: owner, board members, policy makers, facility managers, risk managers, and those who will implement the programs.

3.1.28 Federal Emergency Management Agency. Incremental Seismic Rehabilitation of Retail Buildings. FEMA 399. Washington, DC: Federal Emergency Management Agency, June 2004.
URL: http://www.fema.gov/plan/prevent/rms/index.shtm

FEMA 399: *Incremental Seismic Rehabilitation of Retail Buildings* is a manual that provides owners of retail buildings the information necessary to implement a seismic rehabilitation program. This program, like FEMA 395 through 398, contains three sections: Part A, Critical Decisions for Earthquake Safety in Retail buildings; Part B, Planning and Managing the Process for Earthquake Risk Reduction in Existing Retail Buildings; Part C, Tools for Implementing Incremental Seismic Rehabilitation in Existing Retail Buildings. Each section targets a different audience: senior executives, board members, policy makers, facility managers, risk managers, financial managers, and those who are responsible for the implementation of a seismic program.

3.1.29 Federal Emergency Management Agency. Incremental Seismic Rehabilitation of School Buildings. FEMA 395. Washington, DC: Federal Emergency Management Agency, January 2003.
URL: http://www.fema.gov/plan/prevent/rms/index.shtm

FEMA 395 of the Risk Management Series publication: *Incremental Seismic Rehabilitation for School Buildings* provides school administrators with information to evaluate seismic vulnerability of their buildings and information about a program of seismic rehabilitation for those buildings. The manual consists of three parts: Part A, Critical Decisions for Earthquake Safety in Schools; Part B, Managing the Process for Earthquake Risk Reduction in Existing School Buildings; and Part C, Tools for Implementing Incremental Seismic Rehabilitation in School Buildings.

3.1.30 Federal Emergency Management Agency. Insurance, Finance, and Regulation Primer for Terrorism Risk Management in Buildings. FEMA 429. Washington, DC: Federal Emergency Management Agency, December 2003.
URL: http://www.fema.gov/plan/prevent/rms/index.shtm

This Risk Management Series publication, FEMA 429: *Insurance, Finance, and Regulation Primer for Terrorism Risk Management in Buildings*, is a primer on risk management. The purpose of FEMA 429 is to introduce the building insurance, finance, and regulatory communities to the issue of terrorism risk management in buildings and the tools currently available to manage these risks. It presents current building regulations related to terrorism risk and vulnerability. FEMA 429 also provides a categorized security checklist.

3.1.31 Federal Emergency Management Agency. Primer for Design of Commercial Buildings to Mitigate Terrorist Attacks. FEMA 427. Washington, DC: Federal Emergency Management Agency, December 2003.
URL: http://www.fema.gov/plan/prevent/rms/index.shtm

FEMA 427: *Primer for Design of Commercial Buildings to Mitigate Terrorist Attacks,* a Risk Management Series publication, is a primer that introduces a series of concepts that can help building designers, owners, and managers mitigate the threat of hazards resulting from terrorist attacks on new buildings. FEMA 427 contains extensive qualitative design guidance for limiting or mitigating the effects of terrorist attacks; it focuses primarily on attacks using explosives, but also addresses chemical, biological, and radiological attacks. The primer focuses on limiting the effects of attacks on commercial offices, retail stores, multifamily residences, and light industrial buildings.

3.1.32 Federal Emergency Management Agency. Primer to Design Safe School Projects in Case of Terrorist Attacks. FEMA 428. Washington, DC: Federal Emergency Management Agency, December 2003.
URL: http://www.fema.gov/plan/prevent/rms/index.shtm

FEMA 428: *Primer to Design Safe School Projects in Case of Terrorist Attacks*, from the Risk Management Series publications, is a primer on school projects. The purpose of FEMA 428 is to provide the design community and school administrators with the basic principles and techniques to make a school that is safe from terrorist attacks. It includes a method of risk assessment, information on building design, and school safety plans. It describes the idea of safe rooms that can resist chemical, biological, and radiological attacks as well as blast threats.

3.1.33 Federal Emergency Management Agency. Reference Manual to Mitigate Potential Terrorist Attacks against Buildings. FEMA 426. Washington, DC: Federal Emergency Management Agency, December 2003.
URL: http://www.fema.gov/plan/prevent/rms/index.shtm

This Risk Management Series publication, FEMA 426: *Reference Manual to Mitigate Potential Terrorist Attacks against Buildings* is a reference tool that provides guidance to architects and engineers on how to reduce physical damage to buildings, related infrastructure, and injuries caused by terrorist attacks. The manual presents incremental approaches that can be implemented over time to decrease the vulnerability of buildings to terrorist threats. FEMA 426 is a useful resource that draws information from FEMA, the Department of Commerce, Department of Defense, Department of Justice, General Services Administration, Department of Veterans Affairs, and other organizations.

3.1.34 Federal Emergency Management Agency. Risk Assessment: A How-To Guide to Mitigate Potential Terrorist Attacks against Buildings. FEMA 452. Washington, DC: Federal Emergency Management Agency, January 2005.
URL: http://www.fema.gov/plan/prevent/rms/index.shtm

FEMA 452, *Risk Assessment: a How-To Guide to Mitigate Potential Terrorist Attacks against Buildings* of the FEMA Risk Management Series publications provides a clear, flexible, and comprehensive methodology for preparing a risk assessment. FEMA 452 outlines methods for identifying the critical assets and functions within buildings, determining the threats to those assets, and assessing the vulnerabilities associated with those threats. The guide presents five steps and multiple tasks within each step that define a process for conducting a risk assessment and for selecting risk mitigation strategies.

3.1.35 Federal Facilities Council. Key Performance Indicators for Federal Facilities Portfolios. Federal Facilities Council Technical Report #147. Washington DC: National Academies Press, 2004.

This report discusses the performance measures for federal buildings needed to make informed decisions about facility investment. It outlines the information that answers the following questions: what facilities are present, what is their current condition, what facilities are needed, what issues need to be addressed, what are the required investments, and what are the current investments.

3.1.36 Frangopol, M. Dan. Bridge Safety and Reliability. Reston, VA: American Society of Civil Engineers, 1999.

Bridge Safety and Reliability discusses "bridge reliability concepts and methods, modeling of bridge loads, bridge strength, reliability of bridge components and systems, bridge evaluation, nondestructive bridge testing, bridge code calibration, bridge management systems, and life-cycle cost analysis for bridges." This text is targeted toward engineers, consultants, and those involved in research and development.

3.1.37 Garcia, Mary Lynn. The Design and Evaluation of Physical Protection Systems. Boston, MA: Butterworth-Heinemann, 2001.

This text book is a guide to security system design and integration. It emphasizes "component performance measures" to establish security effectiveness and provides a process for estimating security performance against threats. The contents of this text include the design and evaluation of physical security systems, target identification, alarm assessment, and risk assessment. The fundamental principles presented in this text were the base for the security risk assessment methodologies at Sandia National Laboratories (see reference 2.2.16 on page 19).

3.1.38 Grance, Tim, Joan Hash, Marc Stevens, Kristofor O'Neal, Nadya Bartol, and Robert Young. Guide to Information Technology Security Services: Recommendations of the National Institute of Standards and Technology. Special Publication 800-35. Gaithersburg, MD: National Institute of Standards and Technology, October 2002.
URL: http://csrc.nist.gov/publications/nistpubs/800-35/NIST_SP800-35.pdf

This NIST publication is a guide to selecting, implementing, and managing IT security services. It discusses six phases of IT security services life cycle. Phase 1 evaluates the need for investigating an IT security service and determines if it will improve their security program. In Phase 2 the organization evaluates the current environment and identifies viable solutions. Phase 3 is the decision process, and Phase 4 is the implementation process. Phase 5 ensures that the operation is successful, and phase 6 is the end of the IT service.

3.1.39 Grossi, Patricia, Howard Kunreuther, and Chandu C. Patel. Catastrophe Modeling: A New Approach to Managing Risk. New York, NY: Springer Science and Business Media, 2005.

This text uses the knowledge of a number of experts to analyze how catastrophe models can be used to assess and manage the risks of natural disasters. Three leading catastrophe modeling firms were involved in the development of this book: AIR Worldwide (see reference 2.2.2 on page 15), EQECAT (see reference 3.2.7 on page 57), and Risk Management Solutions (see reference 2.1.22 on page 14). Dennis Kuzak and Tom Larsen from EQECAT present models for insurance rates. It includes discussions on actuarial principles, catastrophe models, differentiation of risk, and rate setting procedures. Weimin Dong and Patricia Grossi from Risk Management Solutions present a section on insurance portfolio management. It discusses portfolio composition, exceedance probability, uncertainty, and risk. David Lalonde from Air Worldwide presents a section on risk financing that includes discussions on probable maximum loss, level of risk, risk transfer, and the cost of risk transfer. There are several other sections presented in this text that include discussions on catastrophe models and insurance, natural hazard risk assessment, uncertainty, risk financing, risk management strategies, and risk transfer. The final chapter discusses the use of catastrophe modeling being applied to terrorism.

3.1.40 Hicks and Associates Incorporated. National Technology Plan for Emergency Response to Catastrophic Terrorism. National Memorial Institute for the Prevention of Terrorism, April 2004.
URL: http://www.mipt.org/pdf/Project-Responder-National-Technology-Plan.pdf

This document published by MIPT (see reference 3.3.2.11 on page 67) focuses on "preventing and deterring terrorism or mitigating its effects." It focuses on technology development in response to terrorist risks. Topics include the following items: personal protection and equipment; detection, identification and assessment; response and recovery; communications; emergency management preparations and planning; biological events; medical response; criminal investigations; and mitigation. The publication discusses the capabilities, limitations, and goals for each topic. It also discusses research and current projects that are being conducted on each topic.

3.1.41 Institute for Business and Home Safety.
URL: http://www.ibhs.org

Dedicated to reducing deaths, injuries, property damage, and economic loss, the Institute for Business & Home Safety (IBHS) produces numerous publications. They address home safety issues involving earthquakes, floods, hail, tornados, and other disasters. In 2005, the Institute for Business & Home Safety published *Open for Business* as a guide for business owners to prepare for and react to disasters. The IBHS document notes that the threat of a disaster-related closure is especially great for small and mid-sized businesses because they lack the financial resources for recovery and risk mitigation, ready access to alternative suppliers, and other advantages that most large organizations possess. The IBHS website provides lists and check lists for business and home owners to prepare against disasters. It includes the following topics: earthquakes, flood, freezing weather, hail, hurricane, tornado, water damage, wildfire, fortification, maintenance, recovery, land use, business protection, disaster safety, and reports of IBHS. Most of the articles within each topic are two or three page brochures on protecting and preparing for disasters.

3.1.42 International Association of Earthquake Engineering. Regulations for Seismic Design – A world List 2004. Tokyo: International Association for Earthquake Engineering, 2004.
URL: http://www.iaee.or.jp/

The International Association of Earthquake Engineering (IAEE) collaborates researchers of earthquake engineering through an exchange of knowledge and ideas. Their website contains the publication *Regulations for Seismic Design*; it lists earthquake designs for over 50 countries. This document can be purchased through IAEE on CD or in print.

3.1.43 International Code Council.
URL: http://www.iccsafe.org

The International Code Council (ICC) protects the health, safety, and welfare of people through safety and performance standards. The ICC is a nonprofit organization that was established in 1994 by the Building Officials and Code Administrators International, Inc. (BOCA), International Conference of Building Officials (ICBO), and Southern Building Code Congress International, Inc. (SBCCI). It was established in order to standardize the nation's code system, to eliminate redundancy, and to optimize the resources of code officials. It also offers a number of services: code application assistance, educational programs, certification programs, plan reviews, and training. The following publications are produced by the ICC: International Building Code, International Residential Code for One and Two Family Dwellings, International Plumbing Code, International Mechanical Code, International Fire Code, International Property Maintenance Code, International Existing Building Code, International Fuel Gas Code, International Energy Conservation Code, International Wild land Urban Interface Code, International Zoning Code, International Private Sewage Disposal Code, and the ICC Electrical Code Administrative Provisions.

3.1.44 International Strategy for Disaster Reduction (ISDR).
URL: http://www.unisdr.org/

The International Strategy for Disaster Reduction is an entity of the United Nations. Their website contains data, statistics, information, and a library on the effect and occurrence of disasters. The information is categorized by country or type of disaster. The library has an online searchable catalogue, but the texts are not available to view online. Many of the statistics on the website are in graphical form, but the raw data is available from EM-DAT (see resource 2.3.1.28). The ISDR website has an online library that is helpful for finding information and data. It contains a list of bibliographies, acquisitions, and clearinghouses among other things.

3.1.45 Joint Chiefs of Staff. Joint Tactics, Techniques, and Procedures for Antiterrorism. Joint Pub 3-07.2. Joint Chiefs of Staff, March 17, 1998.
URL: http://www.fas.org/irp/doddir/dod/jp3_07_2.pdf

Joint Tactics, Techniques, and Procedures for Antiterrorism delineates national policy and general objectives relating to antiterrorism. The document explains department of defense and U.S. government agency command and control relationships. It provides useful tactics, techniques, and procedures in U.S. antiterrorism operations. The targeted audience for this document is the armed forces. However, it provides useful ideas and information on antiterrorism tactics for the public and private sector.

3.1.46 Journal of Homeland Security and Emergency Management.
URL: http://www.bepress.com/jhsem/

The Journal of Homeland Security and Emergency Management is a publication of the Berkeley Electronic Press. The journal was first published in 2004 and addresses various topics associated with homeland security.

3.1.47 Marshall, Harold E., Robert E. Chapman, and Chi J. Leng. "Risk Mitigation Plan for Optimizing Protection of Constructed Facilities." Cost Engineering. Vol. 46, No. 8 (August 2004) 26.

This article describes a three-step protocol for satisfying the need to optimize protection against natural and man-made hazards. Step 1 is to "assess the risk of uncertain, costly, man-made and natural hazards, including terrorism, floods, earthquakes, and fire." Step 2 includes the identification of risk mitigation strategies that reduce the expected value of damages from such events. In step 3 one evaluates the "life-cycle economic effectiveness" of each of the strategies.

3.1.48 McEntire, David A, Robie Jack Robinson, and Richard T Weber. Managing the Threat of Terrorism. Washington DC: International City/County Management Association, 2002.

This report discusses the actions that communities can take to prevent, prepare, and respond to a terrorist attack. Issues include attention to emergency response, security, protecting the scene for investigation, and management.

3.1.49 Mileti, Dennis. Disasters by Design: A Reassessment of Natural Hazards in the United States. Washington, DC: Joseph Henry Press, 1999.

This text was developed from the second national assessment of natural and related technological hazards and disasters. It discusses the development of a sustainable mitigation plan for natural hazards including land-use management, building codes, insurance, forecasting, and engineering. The text also discusses disaster response and preparedness along with the recovery process. The text is aimed at a general audience that includes policymakers and researchers.

3.1.50 Multidisciplinary Center for Earthquake Engineering Research.
       URL: http://mceer.buffalo.edu/

This organization develops and promotes the use of technology to reduce earthquakes. It provides technical reports of current research, distributes a bulletin, and produces publications on earthquake codes. The website has information on bridges and roads as well as buildings and structures. A yearly technical report of current research is available to subscribers.

3.1.51 Nadel, Barbara A. Building Security: Handbook for Architectural Planning and Design. United States: McGraw Hill, 2004.

This text is a handbook of security measures for constructed facilities. It begins with a discussion of the lessons learned from September 11, 2001 and other major events. Part 2 of the text details the planning and design of different types of facilities and facility issues. Part 3 details the engineering of buildings for security. In Part 4 construction cost estimating and emergency response is discussed. Technology and materials are discussed in Part 5 and codes and liabilities are discussed in Part 6. This text is non-technical in nature.

3.1.52 National Academies: Federal Facilities Council
       URL: http://www7.nationalacademies.org/ffc/index.html

The Federal Facilities Council at the National Academies is a "cooperative association of 25 federal agencies with interests and responsibilities related to all aspects of facility design, acquisition, management, maintenance, and evaluation." A number of reports and publications related to facilities can be purchased through their website.

3.1.53 National Association of Home Builders Research Center. ToolBase Services.
       URL: http://www.toolbase.org

ToolBase Services is a homebuilding website for "technical information on building products, materials, new technologies, business management, and housing systems." Within the site is the PATH Technology Inventory, which is a database of housing technologies. It includes plumbing, HVAC systems, energy efficient lighting, storm resistant roofing, and other technologies. Each section includes a description of a selected set of technologies, their cost, ease of use, energy efficiency, quality/durability, environmental performance, and whether they mitigate against disasters. It also describes the development stage of each technology. Within the ToolBase website are recommendations for homeowners to mitigate natural disasters.

3.1.54 National Capital Planning Commission. Designing for Security in the Nation's Capital. Washington DC: NCPC, October 2001.
URL: http://www.ncpc.gov/planning_init/security/DesigningSec.pdf

This text describes the interim security measures that were necessary following September 11, 2001. It describes design solutions for Pennsylvania Avenue and its impact on historic resources, traffic, and the downtown economy.

3.1.55 National Earthquake Hazards Reduction Program (NEHRP).
URL: http://www.nehrp.gov/

This program was established by the Federal Government to reduce the risk of loss caused by earthquakes. Four agencies are involved in the program: the Federal Emergency Management Agency (FEMA), the National Institute of Standards and Technology (NIST), the National Science Foundation (NSF), and the United States Geological Survey (USGS). The NEHRP website provides a list of resources and a number of publications, which include FEMA 356 (see reference 3.1.9 on page 35) and FEMA 450 (see reference 3.1.12 on page 36). The former is a publication concerning the seismic rehabilitation of buildings, and the later concerns recommended seismic regulations for new buildings and structures.

3.1.56 National Fire Protection Association. Standard on Disaster/Emergency Management and Business Continuity Programs. NFPA 1600. Quincy, MA: National Fire Protection Association, 2004.
URL: http://www.nfpa.org/assets/files/PDF/NFPA1600.pdf

The National Fire Protection Association 2004 edition of NFPA 1600, *Standard on Disaster/Emergency Management and Business Continuity Programs* establishes a common set of criteria for disaster management, emergency management, and business continuity programs. The standard provides those individuals with the responsibility for disaster and emergency management and business continuity programs the criteria to assess current programs or develop a program to prepare for disasters.

3.1.57 National Fire Protection Association and Society of Fire Protection Engineers. SFPE Handbook of Fire Protection Engineering. 3rd edition. Quincy, MA: National Fire Protection Association, 2002.

This text contains hundreds of pages of technically detailed information for the design and performance of fire protection engineering. It discusses mechanics of fluids, conduction of heat, convection heat transfer, radiation heat transfer, thermo chemistry, and structural mechanics among other things. This text is intended for use by building code officials, building owners, fire officials, engineers, and architects.

3.1.58 National Incident Management System (NIMS).
URL: http://www.nimsonline.com/

The National Incident Management System (NIMS) was developed by the Department of Homeland Security and is the first set of standardized processes and procedures for emergency responders. It provides standard organizational structure, training, qualifications, equipment, and management.

3.1.59 National Information Service for Earthquake Engineering.
URL: http://nisee.berkeley.edu/

The National Information Service for Earthquake Engineering provides basic descriptions and photos of structural designs along with descriptions of earthquake phenomena. Topics include ground failure, ground shaking, solutions, foundations, superstructure, construction, related research, and historical earthquakes.

3.1.60 National Institute for Occupational Safety. Guidance for Protecting Building Environments from Airborne Chemical, Biological, or Radiological Attacks. Cincinnati, OH: National Institute for Occupational Safety, 2002.
URL: http://www.cdc.gov/niosh/bldvent/pdfs/2002-139.pdf

This publication provides methods to mitigate and prepare for a chemical, biological, or radiological attack. It discusses physical security, things not to do, ventilation, filtration, maintenance, administration, and training. The target audience includes owners, managers and maintenance personnel of both publicly and privately owned buildings. It does not address high risk facilities such as military or law enforcement facilities.

3.1.61 National Institute of Building Sciences.
URL: http://www.nibs.org/

The National Institute of Building Sciences (NIBS) is an interface between the government and the private sector. Its goals are to "improve the building regulatory environment; facilitate the introduction of new and existing products and technology into the building process; and disseminate nationally recognized technical and regulatory information. NIBS played a role in the development of HAZUS-MH (see reference 2.2.7 on page 16) and provides the Whole Building Design Guide (see reference 3.1.62 on page 49). Their website contains publications on asbestos control, lead hazards, construction metrication, the environment, and school facilities.

3.1.62 National Institute of Building Sciences. "Whole Building Design Guide."
URL: http://www.wbdg.org/

The Whole Building Design Guide is an internet source provided by the National Institute of Building Sciences that provides a significant amount of information on building design. It is organized into two categories-Design Guidance and Project Management. Within Design Guidance there are five subcategories: (1) Building Types;

(2) Space Types; (3) Design Disciplines; (4) Design Objectives; and (5) Products and Systems. There are four subcategories for Project Management: Delivery Teams, Planning and Development, Building Commissioning, and Delivery and Controls. Each subcategory is broken into subjects that have detailed information and lists of resources. It addresses specific types of buildings (e.g., hospitals, youth centers, libraries, schools, office buildings, research facilities, and more) and provides information on codes, standards, and other design criteria.

3.1.63 National Institute of Standards and Technology: Building and Fire Research Laboratory. Best Practices for Project Security. NIST GCR 04-865. July 2004. URL: http://www.bfrl.nist.gov/oae/publications/gcrs/04865.pdf

This NIST report complements *Implementing Project Security Practices* developed by the Construction Industry Institute (see reference 3.1.14 on page 37). It determines the best practices for project security of industrial projects and provides assistance in transitioning from security research to implementation. This report is most appropriate for those who are managing security issues from the very initial phases of building design to the final stages of construction.

3.1.64 National Institute of Standards and Technology: Electronics and Electrical Engineering Laboratory: Office of Law Enforcement Standards. URL: http://www.eeel.nist.gov/oles/

The Office of Law Enforcement Standards (OLES) develops performance standards that ensure the development of safe, dependable, and effective equipment for law enforcement. One issue that OLES is addressing is public safety communication standards. Law enforcement officials and emergency responders face difficult challenges in communicating during a large scale response. Other areas of research include chemical systems, inspection technologies, weapons and protective systems, and forensic science issues.

3.1.65 National Research Council. ISC Security Design Criteria for New Federal Office Buildings and Major Modernization Projects. Washington DC: National Academy Press, 2003.

This text provides expertise in architecture, structural engineering, blast-effects mitigation, mechanical engineering, security, risk management, and other building related knowledge in order to identify "aesthetically appealing architectural solutions that achieve both security and physical protection." This text primarily focuses on blast effects and blast mitigation.

3.1.66 National Research Council. Protecting People and Buildings from Terrorism: Technology Transfer for Blast-Effects Mitigation. Washington DC: National Academy Press, 2001.

This publication discusses the vulnerability of U.S. civilian and military personnel to terrorist attacks. Directed by Congress, the Department of Defense conducted a comprehensive research and testing program to further knowledge on protecting people and buildings from terrorist bomb attacks. The program was called the Blast Mitigation for Structures Program (BMSP). This publication reviews the BMSP program and discusses the dissemination of the knowledge gained from it.

3.1.67 National Research Council: Committee on Science and Technology for Countering Terrorism. Making the Nation Safer: The Role of Science and Technology in Countering Terrorism. Washington DC: National Academy Press, 2002.

This publication provides technical approaches to mitigating vulnerabilities to terrorism. It addresses nuclear, radiological, biological, explosive, and chemical threats to public health, agriculture, infrastructure, transportation, and other systems. For each topic it provides recommendations on how to apply existing technology to mitigate terrorism and how to start research that extends the nations future capabilities.

3.1.68 National Science and Technology Council, Committee on Environment and Natural Resources, Subcommittee on Disaster Reduction. Grand Challenges for Disaster Reduction. Washington, DC: Executive office of the President, June 2005.
   URL: http://www.sdr.gov/workinggroups.html

This publication discusses six challenges to disaster reduction: provide hazard and disaster information; understand the natural processes that produce hazards; develop hazard mitigation strategies and technologies; recognize and reduce vulnerability of interdependent critical infrastructure; assess disaster resilience using standard methods; and promote risk-wise behavior. The document discusses each of these challenges and a framework for action.

3.1.69 Pacific Earthquake Engineering Research Center.
   URL: http://peer.berkeley.edu/index.html

The Pacific Earthquake Engineering Research Center (PEER) is an entity of the National Science Foundation's (NSF) program to reduce earthquake losses. It primarily conducts research on performance engineering to prevent damage caused by earthquakes. PEER has nine institutions each located in a university setting. Their website contains scientific information on material and structural strength and performance in an earthquake. The mother site for NSF is the Engineering Research Center Association (http://www.erc-assoc.org/).

3.1.70 Persily, Andy. "Building Ventilation and Pressurization as a Security Tool." ASHRAE Journal. Vol. 46, No. 9. September 2004.

This article discusses the impact of ventilation on the vulnerability of buildings in the event of a chemical, biological, or radiological attack. Increased indoor air pressure above outdoor air pressure has been advocated by a number of organizations to thwart an outdoor release. This article gives particular attention to the air tightness of the building envelope. Many buildings do not have the air tightness required to be effective against an outdoor release. It discusses the fact that this method is only effective against materials that can be filtered out of the air intake of the ventilation system. Shelter in place is also discussed in this article.

3.1.71 Schiff, Anshel J. Guide to Improved Earthquake Performance of Electric Power Systems. Reston, VA: American Society of Civil Engineers, 1999.

This text discusses methods to improve electric power systems in the event of a seismic event. It discusses at length the failure of high-voltage substations, because this is where most power system damage occurs.

3.1.72 The Infrastructure Security Partnership.
URL: http://www.tisp.org/

The Infrastructure Security Partnership (TISP) is a public-private partnership that promotes collaboration to improve the infrastructure against natural and man-made disasters. TISP was created following September 11, 2001 by eleven organizations. It now has over 180 organizations, and membership is open to U.S. based government, professional, and research organizations. TISP hosts security workshops and an annual congress on "Infrastructure security for the Built Environment (ISBE)." They have a monthly e-newsletter as well as a comprehensive website that contains a list of their publications.

3.1.73 United States Army Corps of Engineers.
URL: http://www.usace.army.mil/

The United States Army Corps of Engineers (USACE) provides engineering for civil works, military facilities, and construction management for various federal agencies. Their website provides useful information for locations with extreme conditions. However, increased security has caused some of the research conducted by the USACE to be classified.

3.1.74 **United States Department of Defense. "DoD Minimum Antiterrorism Standards for Buildings." UFC 4-010-01. United States Department of Defense, July 31, 2002.**
URL: http://www.tisp.org/files/pdf/dodstandards.pdf

On July 31, 2002, the Department of Defense (DoD) published a Uniform Facilities Criteria (UFC), "DoD Minimum Antiterrorism Standards for Buildings." The objective of these criteria is to improve the survival of DoD personnel from terrorist attacks. Although the UFC system applies primarily to military departments, the criteria identify and highlight several key aspects of site planning, structural design, architectural design, and electrical and mechanical design that play a role in protecting buildings from explosives threats. The criteria apply to construction projects beginning in FY 2004, new leases in FY 2006, and lease renewals by FY 2010. They provide an example of explicit tradeoffs between two approaches to improving survival from a terrorist attack on a constructed facility: setback distance and structural hardening. DoD focuses on minimum setback distance as their primary approach which separates it from the General Services Administration (GSA) and the Department of State.

3.1.75 **United States Department of Homeland Security.**
URL: http://www.dhs.gov

The Department of Homeland Security (DHS) is given the responsibility of addressing safe transportation, research, and protection from threats, emergencies, and disasters. DHS has a six point agenda: (1) Increase overall preparedness, particularly for catastrophic events; (2) Create better transportation security systems to move people and cargo more securely and efficiently; (3) Strengthen border security and interior enforcement and reform immigration processes; (4) Enhance information sharing with our partners; (5) Improve DHS financial management, human resource development, procurement and information technology; and (6) Realign the DHS organization to maximize mission performance. DHS sponsored the RAMCAP document discussed in reference 2.1.4 and provides education support for homeland security issues (http://hsdec.org/). It also established the Homeland Security Digital Library, which is a database of publications related to homeland security (see reference in Appendix B: 26 on page 121).

3.1.76 **United States Department of Homeland Security. Guidance on Aligning Strategies with the National Preparedness Goal. July 22, 2005.**
URL: www.ojp.usdoj.gov/odp/docs/StrategyGuidance_22JUL2005.pdf

This publication discusses the implementation of many of the Homeland Security Presidential Directives in Appendix C:3. Topics include the National Incident Management System (see reference 3.1.58 on page 49), National Response Plan (see Appendix C:32), National Infrastructure Protection Plan (see Appendix C:31), information sharing, communication, and weapons detection. The article provides additional Internet sites on many of the topics.

3.1.77 United States Department of Homeland Security: Office of state and Local Government Preparedness. Target Capabilities List. Washington, DC: United States Department of Homeland Security, May 23, 2005.
URL: http://www.ojp.usdoj.gov/odp/docs/TCL1_1.pdf

Homeland Security Presidential Directive 8 established the Secretary of Homeland Security as the principal official coordinating hazard preparedness in the U.S. The secretary gave this responsibility to the Executive Director of the Office of State and Local Government Coordination and Preparedness. From these proceedings came the *Universal Task List: Version 1.1*. It develops 36 capabilities required to accomplish essential tasks that minimize property damage, injuries, and loss of life along with mitigating property damage. Federal, state, and local governing entities will be expected to develop and maintain these capabilities according to their relative ability.

3.1.78 United States Department of Homeland Security: Office of State and Local Government Coordination and Preparedness. Universal Task List: Version 2.1. United States Department of Homeland Security, May 23, 2005.
URL: www.ojp.usdoj.gov/odp/docs/UTL2_1.pdf

The Department of Homeland Security's Office of state and Local Government Coordination and Preparedness established this Universal Task List (UTL) in order to support the national preparedness goal "to engage federal, state, local, and tribal entities, their private and non-governmental partners, and the general public to achieve and sustain risk-based target levels of capability to prevent, protect against, respond to, and recover from major events in order to minimize the impact on lives, property, and the economy." It does not specify what government entities are responsible for performing these tasks. The UTL is a "living" document that will be revised as each of the tasks are implemented.

3.1.79 United States Department of Homeland Security. "Ready.gov."
URL: http://www.ready.gov

Ready.gov is a website of the Department of Homeland Security that provides information on disaster preparation for American businesses and citizens to lessen the effect of a disaster or emergency. It answers questions such as: where should I go in an earthquake, what should I do in case of a fire, and what equipment should be on hand. Since this site targets private individuals, statistical data is beyond the scope of this resource.

3.1.80 United States General Services Administration. Facilities Standards for the Public Buildings Service. P100. March 2005.
URL: http://www.gsa.gov/Portal/gsa/ep/home.do?tabId=1

This text establishes building standards for new buildings and building alterations in the Public Buildings Service (PBS) of the General Services Administration (GSA). It has "policy and technical criteria to be used in the programming, design, and documentation of GSA buildings."

3.1.81 United States Government Accountability Office. "Homeland Security: Actions Needed to Better Protect National Icons and Federal Office Buildings from Terrorism." GAO-050790. Washington, DC: United States Government Accountability Office, June 2005.
URL: www.gao.gov/new.items/d05790.pdf

This government publication, which was produced by the Government Accountability Office (GAO), identifies and discusses the challenges that the Department of the Interior faces in protecting national icons and monuments from terrorism while maintaining public access. This publication also identifies the challenges that the General Services Administration (GSA) faces in protecting federal office buildings that it owns or leases. Actions that the GSA has taken are discussed along with further recommendations. GAO recommended that the Department of the Interior "link the results of its risk assessments and related risk rankings to its funding priorities and also to develop guiding principles for balancing security initiatives with Interior's core mission." It recommended that GSA "establish a mechanism-such as a chief security officer position or formal point of contact-so it is better equipped to address security related matters." This document can be accessed through the GAO: http://www.gao.gov.

3.2 Risk Management Software

3.2.1 AIR.
URL: http://www.air-worldwide.com

See reference 2.2.2 (page 15)

3.2.2 Autodesk.
URL: http://usa.autodesk.com

Although Autodesk is not customized for disaster mitigation applications, its features are suitable for formulating and designing risk mitigation strategies. It provides several software products: Autodesk Inventor, Autodesk AutoCAD Revit Series, Autodesk DWF Composer, AutoCAD, Autodesk Map3D, and Autodesk MotionBuilder. These products are useful resources for designing, managing, and constructing buildings and infrastructure. They provide tools for computer animation, CAD (computer-aided design) and GIS (geographic information system) integration, and the coordination of construction documentation. Autodesk is the creator of the DWG file format, which is commonly used in other CAD software.

3.2.3 BALFOUR Technologies LLC. FourDscape.
URL: http://www.bal4.com/

Although FourDscape is not customized for disaster mitigation applications, its features are suitable for formulating and designing risk mitigation strategies. This software package enables users to "visually integrate and analyze multi-dimensional data flows in a collaborative environment." The user can edit and view a 3D construction project as it develops over time. The FourDscape software has been used for design coordination, construction staging, and project and public communications to analyze airports, traffic, and transit systems.

3.2.4 Bentley Systems. MicroStation.
URL: http://www.bentley.com

Although MicroStation is not customized for disaster mitigation applications, its features are suitable for formulating and designing risk mitigation strategies. It is a software package produced by Bentley Systems that is used by a number of transportation departments for the design, construction, and operation of infrastructural entities. It provides 3D modeling, workgroup productivity development, design animations, and supports the DGN and DWG file formats. Bentley Systems also produces ProjectWise, which coordinates information and data in order to reduce costs and shorten project schedules.

3.2.5 BMS Solutions.
URL: http://www.bmssolutions.com/

Although BMS Solutions is not customized for disaster mitigation applications, its software features are suitable for formulating and designing risk mitigation strategies. It provides management software for financial, manufacturing, and government entities in order to manage health, safety, environmental, and risk issues.

3.2.6 Defense Threat Reduction Agency (DTRA). Consequences Assessment Tool Set (CATS).
URL: http://cats.saic.com/index.html OR
http://www.dtra.mil/press_resources/fact_sheets/fs_includes/cats.jace.cfm

The Consequences Assessment Tool Set (CATS) includes a consequence management tool that combines hazard prediction, consequence analysis, management tools, the Hazard Prediction and Assessment Capability (HPAC) system, and population and infrastructure data in a GIS system. The software uses real-time weather data and other databases to assess the effects of natural disasters as well as man-made hazards. For example, it can predict tidal surges, contamination trajectory, and earthquake damage. It can provide optimal roadblock locations, create scenarios for training and planning, and create contingency plans with its population and infrastructure data. Similar to FEMA's HAZUS software, this program requires ESRI's Arc GIS software for mapping. The user manual is available online at https://www.hsdl.org/homesec/docs/dtic/ADA423521.pdf.

and the program is currently available to federal, state, and local government organizations. CATS along with other DTRA add in software are among the primary modeling software products used by emergency responders and other government entities to model hazards. The Defense Threat Reduction Agency (DTRA) offers training courses for CATS and other software through the Assessment of Catastrophic Events Center (ACECenter): https://acecenter.cnttr.dtra.mil/acecenter/_login.cfm. One must apply for a password to access their site.

3.2.7 EQECAT.
URL: http://www.eqecat.com/

EQECAT provides risk management software for the insurance industry. WORLDCATenterprise is their primary product that provides hazard coverage for 88 countries and allows insurers to assess natural hazard risk exposure.

3.2.8 Intergraph. SmartPlant Review.
URL: http://ppm.intergraph.com/visualization/sp_review.asp

Although SmartPlant Review is not customized for disaster mitigation applications, its features are suitable for formulating and designing risk mitigation strategies. It is a software product that provides a 3D visualization tool for the review of process and power plants during engineering, construction, and maintenance.

3.2.9 National Institute of Standards and Technology (NIST). Building Life Cycle Cost.
URL: http://www.bfrl.nist.gov/info/software.html

See reference 4.1.3.3 (page 90)

3.2.10 National Institute of Standards and Technology (NIST). Cost Effectiveness Software Tool.
URL: http://www.bfrl.nist.gov/info/software.html

See reference 4.1.3.4 (page 91)

3.2.11 Tec-Com Incorporated. RiskWorld.
URL: http://www.riskworld.com/

See reference 2.2.18 (page 20)

3.3 Guidance Documents for Estimating Costs and Losses

3.3.1 Mitigation Costs

3.3.1.1 American Re, 2006.
URL: http://www.amre.com/

See reference in Appendix B:1 (page 117)

3.3.1.2 Amos, Scott. Skills and Knowledge of Cost Engineering. 5th Edition. Morgantown, WV: AACE International, 2004.

This text is a reference guide on "cost estimating, planning and scheduling, progress and cost control, project management, economic analysis, [and] risk." It is developed by the Association for the Advancement of Cost Engineering (AACE), which publishes the monthly *Cost Engineering Journal*. It provides certification as a Certified Cost Engineer (CCE) or Certified Cost Consultant (CCC), which is accredited by the Council of Engineering and Scientific Specialty Boards (CESB).

3.3.1.3 Association for the Advancement of Cost Engineering.
URL: http://www.aacei.org/

The Association for the Advancement of Cost Engineering (AACE) publishes the monthly Cost Engineering Journal and provides certification as a Certified Cost Engineer (CCE) or Certified Cost Consultant (CCC). The CCE and CCC are accredited by the Council of Engineering and Scientific Specialty Boards (CESB).

3.3.1.4 Bledsoe, John D. Successful Estimating Methods. Kensington, MA: RS Means Company, 1992.

This guide is a tool to improve analysis skills of claims, life cycle costs, and uncertainty evaluation. It discusses the basics of estimating and issues beyond the basics. Some of the topics include types of estimates, tools, data sources, methods, "ballpark" estimates, square foot estimates, conceptual estimates, automated estimates, cycle time analysis, and life cycle cost analysis.

3.3.1.5 BOMA International. Experience Exchange Report. Washington, DC: BOMA International, 2006.
URL: http://www.boma.org/

This publication provides benchmarking income and expense data for commercial buildings in a number of U.S. cities. Figures in this publication include administrative expenses, building hours, cleaning expenses, parking income, office area income, occupancy, rentable square feet, repairs and maintenance, road expenses, security expenses, square foot per office worker, operating expenses, and year-end rent. Experience Exchange Report can be purchased in print or electronic version from the BOMA International website.

3.3.1.6   Construction Management Economics.
   URL: http://www.routledge.com/

This refereed journal is published monthly by Routledge and contains research on the management and economics of building and civil engineering. It can be purchased online monthly in electronic format or quarterly in a hard copy.

3.3.1.7   Design Cost Data. Monterey, CA: ISI Publications.
   URL: http://www.dcd.com/

This bimonthly periodical provides construction costs for cost estimating. Their website provides a National Historical Building Cost Database to its subscribers, which contains 1 200 actual projects with the costs broken down into CSI MasterFormat.

3.3.1.8   Engineering News Record.
   URL: http://www.enr.com/

The Engineering News Record (ENR) provides business and technical news on the construction industry. It discusses and provides data on new technology, markets, costs, regulations, equipment, materials, and covers construction news from around the world. Articles are available for purchase through their website or free to ENR members.

3.3.1.9   Federal Emergency Management Agency. Second Report on Costs and Benefits of Natural Hazard Mitigation. Washington, DC: Federal Emergency Management Agency, August 1998.
   URL: http://www.fema.gov/pdf/library/haz_pbo.pdf

This FEMA publication contains case studies of organizations that benefited from natural hazard mitigation. Warner Brothers Studios, BellSouth, Hewlett-Packard Company, and other major businesses conducted natural hazard mitigation and saved millions of dollars after enduring a natural hazard.

3.3.1.10 Federal Highway Administration. Meeting the Customer's Needs for Mobility and Safety during Construction and Maintenance Operations. HPQ – 98 – 1. Washington, DC: United States Department of Transportation, September 1998.
   URL: http://www.fhwa.dot.gov/reports/bestprac.pdf

An important concern for construction is the redirection of traffic; congestion costs were estimated to be $51 billion in 1993. This publication discusses the safe and efficient flow of traffic through construction zones, which can have a significant impact on a local economy.

3.3.1.11 Frank R. Walker Company. Walker's Building Estimator's Reference Book. Chicago, IL: Frank R. Walker Company.
    URL: http://www.frankrwalker.com/

This text provides building costs, labor productivity, and material quantities along with the tools needed to make construction cost estimates. It is organized in CSI format (see MasterFormat in reference 4.1.2.30) and discusses construction techniques, types of material, and application rates.

3.3.1.12 Godschalk, David R., Timothy Beatley, Philip Berke, David J. Brower, Edward J. Kaiser, Charles C. Bohl, and R. Matthew Goebel. Natural Hazard Mitigation. Washington, DC: Island Press, 1999.

See reference Appendix C:7 (page 124)

3.3.1.13 International Trade Administration. U.S. Industry & Trade Outlook.
    URL: http://www.ita.doc.gov/td/industry/otea/outlook/index.html

The International Trade Administration (ITA) is a part of the Department of Commerce. It provides information on international markets, protects the U.S. from unfair competition, and ensures access to international markets. The *U.S. Industry & Trade Outlook* is an industry-by-industry overview of the U.S. economy. It provides historical data on shipments, imports, exports, and employment along with industry trends, technology, and international competition. This publication provides one, two, and five year forecasts of each industry.

3.3.1.14 Lufkin, Peter S., and Robert M. Silsbee. The Whitestone Building Maintenance and Repair Cost Reference 2005-2006. Seattle, WA: Whitestone Research, 2005.

This text is a source of building maintenance and repair cost statistics. It provides costs to maintain a building over its service life, inflation of construction costs, location cost adjustments, and information on the expected lifetime of a building. This text also provides a 50-year cost profiles for individual buildings.

3.3.1.15 Munich Re Group.
    URL: http://www.munichre.com/

See reference 3.3.2.10 (page 66)

3.3.1.16 Owen, David D. Building Security: Strategies & Costs. Kingston, MA: Reed Construction Data, 2003.

In 2003, RS Means published *Building Security: Strategies & Costs* as a text to assist building owners and managers assess risk and vulnerability to their buildings, develop emergency response plans, and make choices about protective measures and designs.

Building Security also includes pricing information for multiple security-related components, systems, and equipment, as well as the labor required for installation. In addition to materials and equipment, the cost data also includes information about other security and prevention measures such as command (guard) dogs, exterior plants, and planters.

3.3.1.17 Ritz, George J. Total Construction Project Management. United States: McGraw-Hill, 1994.

Total construction Project Management provides information on preparing bids and proposals, project planning, scheduling, budgeting, and organization. This text presents a "total systems approach" to construction management. It also discusses construction safety and health, communications, and the application of computers.

3.3.1.18 RS Means.
URL: http://www.rsmeans.com/

RS Means provides a significant amount of cost data for buildings and other structures, which can be used to estimate project costs. Their publications include the following titles: *Mechanical Cost Data, Plumbing Cost Data, Electrical Cost Data, Facilities Maintenance & Repair Cost Data, Repair and Remodeling Cost Data, Facilities Construction Cost Data, Residential Cost Data, Commercial Cost Data, Site Work & Landscape Cost Data, Open Shop Building Construction Cost Data, Heavy Construction Cost Data, Interior Cost Data, Concrete & Masonry Cost Data, Labor Rates for the Construction Industry,* and *Means City Cost Data.* These publications can be ordered on RS Means' website: http://www.rsmeans.com/. RS Means also provides a software program that allows the user to view and automatically update data (see reference 3.4.7 on page 72).

3.3.1.19 RS Means. Building Construction Cost Data. Kingston, MA: RS Means.

Building Construction Cost Data provides a resource of cost data for individual costs of material, labor, equipment, and a crew associated with building construction. It provides historical cost indexes and location factors that affect the cost of building construction. The information on current costs is detailed to a particular type of a certain material. For instance, the cost of a particular type of concrete material or the cost of loading dock equipment is included in the data.

3.3.1.20 RS Means. Means Square Foot Costs. Kingston, MA. RS Means.

This publication provides square foot costs for 100 structures and for thousands of modifications that can be used to make quick cost estimations. There is both a residential and commercial section. The residential section provides costs for four classes of construction in seven building types. The commercial section provides costs for 72 model buildings. In a separate section, costs are broken into "assemblies" for more detailed estimates. The assemblies section is broken into substructure, shell, interiors, services, equipment and furnishings, special construction, and building site work.

3.3.1.21 RS Means. Unit Price Estimating Methods. 3rd Edition.
URL: http://www.rsmeans.com/

This text is a guide for preparing unit cost estimates. It begins by discussing four estimate types: order of magnitude, square foot and cubic foot, assemblies, and unit price estimates. The chapters that follow discuss quantity takeoff, pricing, bid scheduling, computerized estimating, and using Means cost data. Typically, a unit price estimate is the most accurate of the four construction cost estimates.

3.3.1.22 RS Means. Square Foot and Assemblies Estimating Methods. 3rd Edition.
URL: http://www.rsmeans.com/

This text is a guide to making construction cost estimates in a timely manner by using square foot and assemblies cost data. These estimates are considered to be less accurate than a unit price estimate, but require less time and information; square foot and assemblies cost estimates are conceptual or appraisal cost estimates.

3.3.1.23 Swiss Re.
URL: http://www.swissre.com/.

See reference 3.3.2.22 (page 69)

3.3.1.24 United States Department of Commerce: United States Census Bureau.
URL: http://www.census.gov

See reference 2.3.2.15. (page 30)

3.3.1.25 United States Department of Commerce: United States Census Bureau. Construction Statistics.
URL: http://www.census.gov/const/www/

This data is taken on a monthly basis and tracks housing units. It measures the value of all construction put in place each month and the value of residential improvement and repair work each quarter. Additionally, the Census of Construction Industries is taken every five years and provides an overview of the construction industry. The information on this site is categorized by the following topics: new residential construction, new residential sales, construction price indexes, characteristics of new housing, construction spending, residential improvements, manufactured housing, and census of construction.

3.3.1.26 United States Department of Energy.
   URL: http://www.doe.gov

The Department of Energy (DOE) provides energy efficiency and security information while protecting the environment and advancing the economy. Within the DOE, the Office of Energy Efficiency and Renewable Energy (EERE) leads the government's research, development, and deployment of energy efficiency. It invests in valuable research that, because of high risks, would not otherwise be conducted. It provides information on energy efficiency in homes, cars, buildings, and industry.

3.3.1.27 United States Department of Energy: Energy Efficiency and Renewable Energy.
   URL: http://www.eere.energy.gov/

This Energy Efficiency and Renewable Energy website is a web portal to relevant online resources. It provides energy efficiency information on buildings, transportation, and industry and discusses renewable energy such as biomass, geothermal, hydrogen, solar, and wind energy.

3.3.1.28 United States Department of Energy: Energy Information Administration.
   URL: http://www.eia.doe.gov/

The Energy Information Administration (EIA), created by Congress in 1977, is an agency of the U.S. Department of Energy that provides statistical information. The EIA collects independent energy data for reliable policy making, and public understanding of energy and its interaction with the economy and environment. The consumers of EIA data include congressional, government, industrial, academic, financial, media, and public institutions. The data provided is searchable by type of fuel (i.e., petroleum, natural gas, electricity, coal), by topic (i.e., consumption, forecasts, environment), and by type of report (i.e., brochures, service reports, survey forms). *Gasoline and Diesel Fuel Update, International Energy Outlook, Residential Energy Consumption Survey,* and the *Monthly Energy Review* are among the many reports of the EIA.

3.3.1.29 United States Department of Energy: Energy Information Administration, Annual Energy Outlook.
   URL: http://www.eia.doe.gov/oiaf/aeo/

The *Annual Energy Outlook (AEO)* offers an outlook on energy through 2025. Using the EIA National Energy Modeling System, AEO provides useful data for life cycle cost analysis. There are numerous forecast tables along with information on energy prices, economic growth, energy trends, energy consumption, energy intensity, electricity generation, emissions, and energy imports.

3.3.1.30 United States Department of Energy: Energy Information Administration. Annual Energy Review.
URL: http://www.eia.doe.gov/aer

Published by the EIA, the *Annual Energy Review (AER)* provides current and historical annual data from 1949 to the present. It includes total energy production, consumption, and trade for the following energy sources: petroleum, natural gas, coal, electricity, and nuclear energy.

3.3.1.31 United States Department of Energy: Energy Information Administration, Commercial Buildings Energy Consumption Survey.
URL: http://www.eia.doe.gov/emeu/cbecs/

The *Commercial Buildings Energy Consumption Survey (CBECS)* is quadrennial survey started in 1979 taken by the Energy Information Administration. It is a national survey of the stock of U.S. commercial buildings, energy characteristics, and energy consumption and expenditure. Any building with at least half of the floor space reserved for a purpose that is not residential, industrial, or agricultural is considered a commercial building. This definition does include buildings that might not traditionally be considered commercial, such as schools, correctional institutions, and buildings used for religious purposes.

3.3.1.32 United States Department of Energy: Energy Information Administration, Country Analysis Briefs.
URL: http://www.eia.doe.gov/emeu/cabs

The *Country Analysis Briefs* provide a short energy data report for countries around the world. It has information on oil, natural gas, coal, and electricity. It also provides an economic and environmental overview of each nation. Additional links for other information resources are listed on the website.

3.3.1.33 United States Department of Labor: Bureau of Labor Statistics.
URL: http://www.bls.gov

See reference 4.2.24 (page 95)

3.3.1.34 United States Department of Transportation.
URL: http://www.dot.gov/

The U.S. Department of Transportation (DOT) ensures that the nation has a safe and efficient transportation system that meets the changing needs of the U.S. There are thirteen DOT organizations, which include the Federal Highway Administration, Federal Railroad Administration, National Highway Traffic Safety Administration, and Pipeline and Hazardous Materials Safety Administration. These organizations provide transportation information and regulations that can be useful in the construction process. For transportation statistics see reference 2.3.2.18 on page 30.

3.3.2 Event-Related Losses

3.3.2.1 Bureau of Labor Statistics. Monthly Labor Review.
URL: http://www.bls.gov/opub/mlr/mlrhome.htm

The *Monthly Labor Review* is a publication of the Bureau of Labor Statistics that includes relevant labor data and research articles. Topics from past issues include unemployment, state labor legislation, retirement plans, and employment and unemployment statistics. Typically there is a section of articles and a section of reports.

3.3.2.2 Centers for Disease Control, National Center for Health Statistics.
URL: http://www.cdc.gov/nchs/index.htm

The National Center for Health Statistics provides a significant amount of information and statistics on America's health. Their website provides birth data, mortality data, hospital data, health care data, immunization data, and other national health statistics.

3.3.2.3 Environmental Cost Handling Options and Solutions.
URL: http://www.echos-online.com

Environmental Cost Handling Options and Solutions (ECHOS) collects construction and environmental restoration cost information. Data from ECHOS has been used on over 10,000 environmental restoration projects throughout the U.S.

3.3.2.4 Fisk, William J. and Arthur H. Rosenfeld. "Estimates of Improved Productivity and Health from Better Indoor Environments." Indoor Air: 1997. vol. 7. Issue 3. 158-172.

There is mounting evidence that indoor environments significantly affect worker performance. Only rudimentary estimates exist as to what extent productivity increases with increased indoor air quality. This article discusses the evidence and estimation of indoor environmental quality's effect on productivity and health. Estimated U.S. potential savings plus productivity gains from indoor environments are $30 billion to $170 billion.

3.3.2.5 Horowitz, John K., and Richard T. Carson. "Discounting Statistical Lives." Journal of Risk and Uncertainty. Vol. 3, Number 4 (December 1990): 403-413.

This article discusses the discount rate used to calculate the future value of a statistical life for benefit-cost analysis. It uses observations on discrete choices between projects with different time horizons in order to determine a median discount rate. It also suggests that different rates apply to different types of risk.

**3.3.2.6 International Trade Administration. U.S. Industry & Trade Outlook.
URL:** http://www.ita.doc.gov/td/industry/otea/outlook/index.html

See section 3.3.1.13

**3.3.2.7 Kuchler, Fred, and Elise Golan, Assigning Values to Life: Comparing Methods for Valuing Health Risks. Agricultural Economic Report (AER) 784. Washington DC: United States Department of Agriculture, November 1999.
URL:** http://www.ers.usda.gov/Publications/aer784/

This text discusses five methods that economists and analysts use to evaluate policies that affect health and safety: cost-of-illness, willingness-to-pay, cost-effectiveness analysis, risk-risk analysis, and health-health analysis. It explores the assumptions and applications of each method in order to come to four conclusions: (1) the approaches are not interchangeable because they measure different factors; (2) usefulness depends on the unit of measure; (3) most of the approaches integrate income and circumstance; and (3) the theory of willingness-to-pay conflicts with its application.

3.3.2.8 Lister, Debra Brinegar, Elisabeth M. Jenicek, and Paul Fredrick Preissner. Productivity and Indoor Environmental Conditions Research: An Annotated Bibliography for Facility Engineers. USACERL Special Report 98/96. Construction Engineering Research Laboratory, July 1998.

This text provides a list of resources on indoor environments to improve productivity while conserving energy. It contains sources on measuring worker productivity, lighting, thermal control, air quality, and the Hawthorne effect.

3.3.2.9 Moore, Michael J., and W. Kip Viscusi, "The Quantity Adjusted Value of Life." Economic Inquiry: 1998. vol. 26. 369-388.

This article discusses the value of life as it relates to compensation for risk. It also discusses the implicit discount rate that workers use in making life-cycle employment decisions. It estimates that this discount rate ranges between 10 % and 12 % and the implicit value per year of life is $175 000.

**3.3.2.10 Munich Re Group.
URL:** http://www.munichre.com/

The Munich Re Group is a large reinsurance company that has branches and offices throughout the world. It produces a number of annual publications on casualty risk, geo risks, life and health, liability, property and casualty damage, and current topics within the industry.

3.3.2.11 National Memorial Institute for the Prevention of Terrorism (MIPT).
URL: http://www.mipt.org/

The National Memorial Institute for the Prevention of Terrorism (MIPT) located in Oklahoma City is a non-profit corporation established by the surviving families of the Murrah Federal Building bombing in 1995. It has numerous publications and has coordinated with the RAND Corporation in developing the Terrorism Knowledge Base (http://www.tkb.org/Home.jsp). The project has two databases: the RAND Terrorism Chronology Database and the RAND-MIPT Terrorism Incident Database. The former records international terrorist incidents that occurred between 1968 and 1997. The latter records domestic and international terrorist incidents occurring from 1998 to the present. The website includes a great deal of information about terrorist groups, incidents, cases, leaders, their countries, and other issues. See RAND Corporation in reference 2.1.21

3.3.2.12 National Ocean Economics Program.
URL: http://noep.csumb.edu/

The National Ocean Economics Program (NOEP) provides economic information on changes and trends along coastal waters. It has several online databases: ocean sector & industry data, coastal economy sector data, demographics, ocean and coastal economy, non-market data, and U.S. commercial marine fisheries. The NOEP website also has nearly a dozen publications, which include reports on gulf coast hurricane damage, California's ocean economy, and the U.S. coastal economy.

3.3.2.13 National Oceanic and Atmospheric Administration. Storm Data (SD).
URL: http://www5.ncdc.noaa.gov/pubs/publications.html

Storm Data is a monthly publication of the National Oceanic and Atmospheric Administration that provides chronological listing of storm paths, deaths, injuries and property damage organized by state. There is also a database of natural hazard and storm data located at http://www4.ncdc.noaa.gov/cgi-win/wwcgi.dll?wwEvent~Storms. The database is searchable by state, date, and type of event. Each search includes the date, time, magnitude, injuries, property damage, and crop damage caused by each event and all the events together. Data is available for droughts, dust storms, fog, floods, funnel clouds, hail, hurricanes, lightning, oceans, snow, precipitation, extreme temperatures, thunderstorms, tornados, and forest fires.

3.3.2.14 National Oceanic and Atmospheric Administration: Coastal and Ocean Resource Economics.
URL: http://marineeconomics.noaa.gov/welcome.html

The Coastal and Ocean Resource Economics (CORE) program produces research on the social and economic impact of outdoor coastal recreation areas. This research includes the economic impact of coral reefs, beaches, ecological reserves, and other coastal areas that provide public and private goods. CORE intends to provide relevant data sets through their website, but currently they are only available on CD by request.

3.3.2.15 National Oceanic and Atmospheric Administration: Damage Assessment, Remediation, and Restoration Program.
URL: http://www.darrp.noaa.gov/

NOAA's Damage Assessment, Remediation, and Restoration Program (DARRP) provides expertise to assess and restore natural resources affected by hazardous substance release and physical impact. The website provides case studies of previous incidences such as the Kuroshima oil spill, Tenyo Maru oil spill, harbor restoration projects, and river restoration projects. These studies provide descriptions of each incident, damage assessment, and cost of restoration.

3.3.2.16 National Research Council. Stewardship of Federal Facilities: A Proactive Strategy for Managing the Nation's Public Assets. Washington DC: National Academy Press, 1998.

This text addresses the continual deterioration of federal government facilities and the maintenance required for their upkeep. Cost effective repairs and maintenance are deferred to an unspecified future date; though this deferment saves money in the short run, it costs the government a significant amount more in the long run.

3.3.2.17 National Technology Information Service.
URL: http://www.ntis.gov/

The National Technical Information Service is part of the U.S. Department of Commerce and is a central resource for scientific, technical, engineering, and business information. It provides information from over 200 federal agencies on a wide variety of topics, which include health, terrorism response, health costs, and numerous other topics.

3.3.2.18 O'Day, Alan. Weapons of Mass Destruction and Terrorism. Burlington, VT: Ashgate Publishing Limited, 2004.

This text is a collection of 29 articles on weapons and terrorism. The articles are from countries around the world. Topics discussed in this text include the threat of terrorism, bio-terrorism and preparedness, pharmaceutical stockpiles, anthrax, smallpox, nuclear terrorism, chemical weapons, mitigation, medical responses to terrorism, and risk analysis. One article is discussed in reference 2.1.5.

3.3.2.19 Population Reference Bureau.
URL: http://www.prb.org

The Population Reference Bureau contains international statistics on population, health, and the environment. These statistics are available on their website and contain the following categories: country, population trends, education, environment variables, health variables, and reproductive variables. Their website contains articles on countries around the world; topics vary over time and by region.

3.3.2.20 Public Entity Risk Institute.
URL: http://www.riskinstitute.org/

See reference 2.1.20 (page 13)

3.3.2.21 RAND Corporation.
URL: http://www.rand.org/

See reference 2.1.21 (page 13)

3.3.2.22 Swiss Re.
URL: http://www.swissre.com/

Swiss Re is a global reinsurer and is recognized for risk and capital management. It shares much of its research through the Swiss Insurance Training Centre (SITC) and *Sigma*. The latter is a regular publication of research by Swiss Re that can be freely accessed online without charge. Swiss Re lists publications by topic: climate change, natural catastrophes, water, insurance-linked securities, terrorism, liability regimes, and nanotechnology. Each of the publications has a particular focus on relevant trends of risks and capital as it relates to the insurance industry. A sample of publications by Swiss Re include the following articles: "The 20 Worst Catastrophes in Terms of Victims: 2004," "The 20 Most Costly Insurance Losses: 2004," "Earthquake Probability Assessment," "Monthly Economic Outlook," "Quarterly US Property and Casualty," and "Terrorism Risks." Swiss Re also provides an online service called CatNet, which provides catastrophe risk estimates, global natural disaster risk maps, and historic insurance loss data. Users are required to apply for access, but it does not require a fee.

3.3.2.23 United States Department of Commerce: United States Census Bureau.
URL: http://www.census.gov

See section 2.3.2.15 on page 30

3.3.2.24 United States Department of Commerce: United States Census Bureau. Statistical Abstract of the United States.
URL: http://www.census.gov/statab/www/

See section 2.3.2.16 on page 30

3.3.2.25 United States Department of Justice. April 3, 2006.
URL: http://www.usdoj.gov/

The Department of Justice (DOJ) enforces the law and defends the interests of the United States against foreign and domestic threats. The DOJ also provides numerous publications on crime, terrorism, and other threats to the public. The most recent terrorism report can be found on their website: http://www.fbi.gov/publications/terror/terroris.htm.

3.3.2.26 United States Department of Transportation: Bureau of Transportation Statistics.
URL: http://www.bts.gov/

See reference 2.3.2.18 (page 30)

3.3.2.27 United States Department of Transportation: National Highway Traffic Safety Administration: National Center for Statistics and Analysis. "Fatality Analysis Reporting System."
URL: http://www.fars.nhtsa.dot.gov/

The Fatality Analysis Reporting System (FARS) is a web-based encyclopedia provided by the National Highway Traffic Safety Administration that provides statistics on transportation fatalities. It provides transportation fatality trends by type of vehicle, time of day, pedestrian involvement, cyclist involvement, and by circumstance. Several traffic safety reports can be accessed from this website that deal with alcohol, children, large trucks, motorcycles, the elderly, cyclists, pedestrians, school transportation, speeding, and young drivers.

3.3.2.28 Viscusi, W. Kip. Fatal Tradeoffs: Public and Private Responsibilities for Risk. New York: Oxford University Press, 1992.

This text is drawn from a collection of Viscusi's writings. He is one of the leading economists in the study of accident risk. The text discusses market valuations of human life, liability, prospective reference theory, the Occupational Safety and Health Administration (OSHA), and the Consumer Product Safety Commission (CPSC). This text is useful for conducting research on the value of life and risk.

3.3.2.29 Viscusi, W. Kip, and Joseph E. Aldy. "The Value of a Statistical Life: A Critical Review of Market Estimates throughout the World." The Journal of Risk and Uncertainty, Vol. 27, Number 1, (2003): 5-76.

The value of a statistical life is related to the tradeoff of between resources and fatality risks. This journal article reviews over 60 studies from 10 different countries on the value of a statistical life. It discusses risk premiums and injury risk premiums along with the role of unionization, the effect of age on the value of a statistical life, and several econometric issues. The article calculates an income elasticity of a statistical life and discusses policy applications related to the value of a statistical life. A working paper copy is available online at NBER's website: http://www.nber.org/papers/w9487.

**3.3.2.30 White House. "Ch. 24 Ranking Regulatory Investments in Public Health."
Analytical Perspectives: Budget of the United States Government.
Washington, DC: United States Government Printing Office, 2003.**
URL: www.whitehouse.gov/omb/budget/fy2003/pdf/spec.pdf

Chapter 24 of this White House publication, which is titled "Ranking Regulatory Investments in Public Health," discusses optimizing the allocation of resources for risk management. It illustrates the use of cost-effectiveness ratios to compare payoffs from regulatory investments. Additionally, it discusses the benefits and limitations of cost-effectiveness analysis as a method for policy decisions.

3.3.2.31 Woo, Gordon. "The Evolution of Terrorism Risk Modeling." Journal of Reinsurance. April 22, 2003.
URL: http://www.rms.com/Publications/EvolutionTerRiskMod_Woo_Journal Re.pdf

See reference 2.1.25 (page 14)

3.4 Software for Estimating Costs and Losses

3.4.1 4Clicks-Solutions.
URL: http://www.4-clicks.com

4Clicks-Solutions develops estimating, project management, quality control, and accounting software. It provides solutions for the government contract method known as job order contracting. It utilizes RS Mean's data and other resources.

3.4.2 Bid2Win.
URL: http://www.bid2win.com/

Bid2Win is a construction estimating software program for infrastructure construction and related work. Since it is compatible with Department of Transportation (DOT) import files, the user can download a project from the DOT and import it directly into Bid2Win. This software also provides error checking and Microsoft compatibility.

3.4.3 Construction Management Software. ProEst.
URL: http://www.proest.com/

Construction Management Software designed ProEst for general contractors, residential builders, remodelers, concrete and masonry contractors, and landscapers. It contains a database of thousands of standard industry items, but also allows for user-defined data. It is organized in CSI format and has access to the RS Means trade databases. The software has a digitizer interface that automatically counts items from blueprints.

3.4.4 HardDollar.
URL: http://www.harddollar.com/

HardDollar is a construction cost estimating software program that is useful for estimating the construction cost of infrastructure such as civil engineering, transportation, water facilities, and environmental projects by providing the required information and tools. This software can track the costs of previous projects, changes in current projects, and profitability.

3.4.5 MC^2.
URL: http://www.mc2-ice.com/

MC^2 develops cost estimating software for construction firms. Their software assists in estimating construction work, earthwork, section work, and trench work. ICE 2000 is their primary estimating program, which contains a cost database for general construction, site work, steel fabrication, steel erection, wastewater tanks, equipment, plumbing, fire protection, HVAC systems, electrical systems, and piping.

3.4.6 Quest Solutions. Quest Estimator.
URL: http://www.questsolutions.com

Quest Estimator is a construction cost estimating software program that contains a digitizer for counting items, a database for cost items, and provides cost estimates. The company offers job specific software for earthwork, trench work, and roadwork with graphical interfaces.

3.4.7 RS Means. CostWorks.
URL: http://www.rsmeans.com

CostWorks is a software program developed by RS Means. It allows the user to download data from RS Mean's website and use it in a spreadsheet or export it to Microsoft Excel in order to make cost estimates. A limited amount of cost estimating can be conducted on their website without purchasing any products. Also see reference 3.3.1.18 on page 61.

3.4.8 Sage Timberline Office.
URL: http://www.sagetimberlineoffice.com/

Sage Timberline Office produces construction cost estimating software, which is recognized throughout the construction industry as "Timberline." It allows the user to either enter personalized cost data or access online data from RS Means in order to make cost estimates. Timberline integrates with other applications and users can add a digitizer or a Computer Animated Drawing (CAD) Integrator to generate quick accurate estimates. The CAD Integrator uses a standard language to define CAD objects such as windows and doors. This standard language allows the software to count the objects within a CAD drawing rather than an engineer counting them by hand.

3.4.9 TreeAge. TreeAge Pro Suite.
   URL: http://www.treeage.com/

TreeAge Pro Suite is a decision analysis software program. It uses decision trees in order to conduct Monte Carlo Simulation, expected value, and sensitivity analysis as well as other decision analyses. Currently TreeAge offers a free trial of their software, which has a help file that walks the user through the program.

3.4.10 U.S. COST. Success Estimator.
   URL: http://www.uscost.com

Success Estimator, developed by U.S. COST, is a construction cost estimating tool. It has thousands of cost items, a city index tool, and allows the use of RS Mean's data or personalized data.

3.4.11 Vertigraph Incorporation. BidPoint.
   URL: http://www.vertigraph.com/

Vertigraph provides software solutions that "automate the takeoff, estimating, bidding and digitizing functions" of construction companies. BidPoint is an Excel add-in program developed by Vertigraph. It allows estimators to digitize, edit, and format drawings and quantities. BidScreen is another Vertigraph excel add-in software product designed for plans that are already in digital format.

3.4.12 WinEstimator Incorporated. WinEst.
   URL: http://www.winest.com

WinEst estimating software along with its additional modules provides a tool to estimate the costs of commercial, residential, public, and specialty structures while integrating accounting, scheduling/management, and CAD software solutions. The user of the software can utilize their own cost database, the cost database that comes with the software, or use one of the four third-party cost databases: RS Means, Environmental Cost Handling Options and Solutions (ECHOS) cost database, MCAA, and NECA.

4 Economic Evaluation Guidance

Economic evaluation is the means through which competing alternatives are analyzed and a cost-effective risk mitigation plan is identified. Risk assessment and risk management, on the other hand, evaluate the risks posed by natural and man-made hazards, formulate the alternative combinations of risk mitigation strategies, and provide the associated cost and hazard data needed to compare the competing alternatives. Risk assessment and risk management ensure that the alternative combinations of risk mitigation strategies are formulated so that they can be rigorously analyzed with economic tools.

The economic evaluation step includes the selection of the appropriate measures of economic performance, a rigorous analysis of the alternative combinations of risk mitigation strategies, the identification of the cost-effective risk mitigation plan, and the documentation necessary to support the recommendation of that plan. The economic evaluation step places special emphasis on the treatment of uncertainty on the selection of a cost-effective risk mitigation plan.

Section 4.1 provides references to the economic tools used in performing an economic evaluation. Economic tools include evaluation methods, standards that support and guide the application of those methods, and software for implementing the evaluation methods. Economic evaluation methods have been widely used in project evaluation for decades. A survey of evaluation methods and a comprehensive treatment of how to apply them is given in Ruegg and Marshall. Many evaluation methods have been transformed into industry consensus standards. The compilation of *ASTM Standards on Building Economics* contains more than 20 standards. Evaluation methods fall under the category of ASTM Standard Practices. The compilation also includes ASTM Standard Guides and Classifications, both of which are useful in implementing the Standard Practices. The format used in ASTM Standard Practices includes a "procedure" section that provides step-by-step instructions for calculating measures of economic performance. Finally, a number of software tools are available that produce economic measures consistent with the Standard Practices.

Section 4.2 provides references to economic modeling resources. Documents abstracted in this section cover assumptions that underlie the economic evaluation, the key parameters that define the project (e.g., purpose of the project, length of the study period, and discount rate), the specification of alternatives, choosing the right evaluation method or combination of evaluation methods, and producing the documentation necessary to support the recommendation of the most cost-effective risk mitigation plan. Comprehensive treatments of these issues are provided in Ruegg and Marshall, Fuller and Petersen, the compilation of *ASTM Standards on Building Economics*, and Marshall's audiovisual series.

Section 4.3 provides references to the treatment of uncertainty in an economic evaluation. Investments in long-lived projects, such as the erection of new constructed facilities or additions and alterations to existing constructed facilities, are characterized by uncertainties regarding project life, operation and maintenance costs, revenues, and other factors that affect project economics. To make reliable economic evaluations, treatment of uncertainty is particularly important for projects affected by natural and man-made hazards that occur infrequently, but have significant consequences. Therefore, when developing a cost-effective risk mitigation plan, three distinct levels of analysis are recommended. This "analysis strategy" systematically adds increased detail to the decision-making process. The starting point for conducting an economic evaluation is to do a baseline analysis. In the baseline analysis, all data elements entering into the calculations are fixed. The term baseline analysis is used to denote a complete analysis in all respects but one; it does not address the effects of uncertainty. Sensitivity analysis measures the impact on project outcomes of changing the values of one or more key data elements about which there is uncertainty. Sensitivity analysis can be performed for any measure of economic performance. Therefore, a sensitivity analysis complements the baseline analysis by evaluating the changes in output measures when selected data inputs are allowed to vary about their baseline values. The key advantage of sensitivity analyses is that they are easily constructed and computed and the results are easy to explain and understand. Their disadvantage is that they do not produce results that can be tied to probabilistic levels of significance. Monte Carlo simulation varies a small set of key parameters either singly or in combination according to an experimental design. Associated with each key parameter is a probability distribution function from which values are randomly sampled. The major advantage of the Monte Carlo simulation technique is that it permits the effects of uncertainty to be rigorously analyzed through reference to a derived distribution of project outcome values.

4.1 Economic Tools

4.1.1 Evaluation Methods

4.1.1.1 ASTM International. ASTM Standards on Building Economics, 5th edition. West Conshohocken, PA: ASTM International, 2004.

See reference 4.1.2.2 (page 82)

**4.1.1.2 ASTM International. "Standard Guide for Developing a Cost-Effective Risk Mitigation Plan for New and Existing Constructed Facilities." ASTM E 2506. Annual Book of ASTM Standards: 2006. Vol. 04.12. West Conshohocken, PA: ASTM International.
URL: http://www.astm.org/**

See reference 4.1.2.8 (page 84)

4.1.1.3 ASTM International. "Standard Guide for Selecting Economic Methods for Evaluating Investments in Buildings and Building Systems." ASTM E 1185. Annual Book of ASTM Standards: 2005. Vol. 04.11. West Conshohocken, PA: ASTM International.
URL: http://www.astm.org/

See reference 4.1.2.9 (page 84)

4.1.1.4 ASTM International. "Standard Guide for Selecting Techniques for Treating Uncertainty and Risk in the Economic Evaluation of Buildings and Building Systems." E 1369. ASTM Annual Book of ASTM Standards: 2005. Vol. 04.11. West Conshohocken, PA: ASTM International.
URL: http://www.astm.org/

See reference 4.1.2.10 (page 84)

4.1.1.5 ASTM International. "Standard Guide for Summarizing the Economic Impacts of Building-Related Projects." E 2204. Annual Book of ASTM Standards: 2005. Vol. 04.12. West Conshohocken, PA: ASTM International.
URL: http://www.astm.org/

See reference 4.1.2.11 (page 85)

4.1.1.6 ASTM International. "Standard Practice for Applying Analytical Hierarchy Process (AHP) to Multi-attribute Decision Analysis of Investments Related to Buildings and building Systems." E 1765. Annual Book of ASTM Standards: 2005. Vol. 04.12. West Conshohocken, PA: ASTM International.
URL: http://www.astm.org/

See reference 4.1.2.13 (page 85)

4.1.1.7 ASTM International. "Standard Practice for Measuring Benefit-to-Cost and Savings-to-Investment Ratios for Buildings and Building Systems." E 964. Annual Book of ASTM Standards: 2005. Vol. 04.11. West Conshohocken, PA: ASTM International.
URL: http://www.astm.org/

See reference 4.1.2.15 (page 86)

4.1.1.8 ASTM International. "Standard Practice for Measuring Cost Risk of Buildings and Building Systems." E 1946. Annual Book of ASTM Standards: 2005. Vol. 02.12. West Conshohocken, PA: ASTM International.
URL: http://www.astm.org/

See reference 4.1.2.16 (page 86)

4.1.1.9 ASTM International. "Standard Practice for Measuring Internal Rate of Return and Adjusted Internal Rate of Return for Investments in Buildings and Building Systems." E 1057. Annual Book of ASTM Standards: 2005. Vol. 02.11. West Conshohocken, PA: ASTM International.
URL: http://www.astm.org/

See reference 4.1.2.17 (page 86)

4.1.1.10 ASTM International. "Standard Practice for Measuring Life-Cycle Costs of Buildings and Building Systems." E 917. Annual Book of ASTM Standards: 2005. Vol. 04.11. West Conshohocken, PA: ASTM International.
URL: http://www.astm.org/

See reference 4.1.2.18 (page 86)

4.1.1.11 ASTM International. "Standard Practice for Measuring Net Benefits for Investments in Buildings and Building Systems." E 1074. Annual Book of ASTM Standards: 2005. Vol. 04.11. West Conshohocken, PA: ASTM International.
URL: http://www.astm.org/

See reference 4.1.2.19 (page 87)

4.1.1.12 ASTM International. "Standard Practice for Measuring Payback for Investments in Buildings and Building Systems." E 1121. Annual Book of ASTM Standards: 2005. Vol. 04.11. West Conshohocken, PA: ASTM International.
URL: http://www.astm.org/

See reference 4.1.2.20 (page 87)

4.1.1.13 ASTM International. "Standard Practice for Performing Value Analysis (VA) of Buildings and Building Systems." E 1699. Annual Book of ASTM Standards: 2005. Vol. 04.11. West Conshohocken, PA: ASTM International.
URL: http://www.astm.org/

See reference 4.1.2.23 (page 88)

4.1.1.14 Center for Risk and Economic Analysis of Terrorism Events.
URL: http://www.usc.edu/dept/create

The Center for Risk and Economic Analysis of Terrorism Events (CREATE) is a university center of excellence funded by the Department of Homeland Security in order to "improve our nation's security through the development of advanced models and tools for the evaluation of the risks, costs, and consequences of terrorism." Their website contains information about current events and research.

4.1.1.15 Chapman, Robert E. Applications of Life-Cycle Cost Analysis to Homeland Security Issues in Constructed Facilities: A Case Study. NISTIR 7025. Gaithersburg, MD: National Institute of Standards and Technology, October 2003.
URL: http://www.bfrl.nist.gov/oae/publications/nistirs/7025.pdf

This publication is a life-cycle cost case study on the renovation of a data center. The renovation includes an upgrade of the HVAC system, telecommunications system, and data processing system along with addressing security concerns. The publication systematically explains the life-cycle cost analysis. It explains each equation and then uses them in the case study of the data center.

4.1.1.16 Chapman, Robert E., and Chi J. Leng. Cost-Effective Responses to Terrorist Risks in Constructed Facilities. NISTIR 7073. Gaithersburg, MD: National Institute of Standards and Technology, March 2004.
URL: http://www.bfrl.nist.gov/oae/publications/nistirs/7073.pdf

This NIST publication presents a three-step process in order to develop and identify optimal risk mitigation plans. The decision methodology is based on life-cycle cost and is supported by a voluntary industry standard, ASTM E 917. This method and a software program to compliment it are detailed within the publication. Other evaluation methods such as present value of net savings, savings to investment ratio, and adjusted internal rate of return are explained and discussed.

4.1.1.17 Environmental Protection Agency (EPA). Guidelines for Preparing Economic Analyses. EPA 240-R-00-003. Washington, DC: United States Environmental Protection Agency, September 2000.
URL: http://www.epa.gov/opei/pubsinfo.htm

Guidelines for Preparing Economic Analyses is part of an effort by the EPA to improve the use of sound science in decision processes and provide guidance for the analysis of the economic impact of policies and regulations. It discusses social and private discounting, costs, benefit valuation, social costs, policies, consumer surplus, value of human health, ecological benefits, and social cost. This EPA publication addresses economic issues that are useful for mitigation assessment with an environmental aspect to it. The EPA has numerous other publications and data that relate to the costs and benefits of an improved environment. They are available through the National Center for Environmental Economics website:
http://yosemite.epa.gov/ee/epa/eed.nsf/webpages/Publications.html

**4.1.1.18 Fuller, Sieglinde K., and Stephen R. Petersen. Life-Cycle Costing Manual for the Federal Energy Management Program. Hand Book 135, 1995 Edition. Gaithersburg, MD: National Institute of Standards and Technology, 1995.
ULR:** http://fire.nist.gov/bfrlpubs/build96/PDF/b96121.pdf

This manual is a guide to the Life Cycle Cost (LCC) method for economic evaluation. It expands the methods and criteria of the Federal Energy Management Program (FEMP) rules published in 10 CFR 436, Subpart A. The manual provides an explanation of the life-cycle cost (LCC) method, defines the measures of economic performance, describes assumptions that are made, and gives examples of the methods. Life Cycle Cost Analysis (LCCA) uses all the costs pertaining to owning, operating, maintaining, and disposing of a project to evaluate whether it is an economical project. This method can be used to evaluate any investment decision that has higher investment costs with future savings and is particularly useful when considering energy saving investment projects. Other topics that are explained and discussed include the discount rate, price escalation, net savings, savings-to-investment ratio, adjusted internal rate of return, discounted payback, and simple payback.

**4.1.1.19 Mackin, T.J., LTC Darrall Henderson, and J.W. Jones. A Method For Allocating Financial Resources to Combat Terrorism: Optimizing the Reduction of Consequences. 71^{ft} MORS Symposium, Working Group 16, June 13, 2003. Alexandria, VA: Military Operations Research Society, October 2003.
URL:** http://handle.dtic.mil/100.2/ADA418274

See reference 4.2.16 (page 94)

4.1.1.20 Marshall, Harold E. Audiovisual series on Least-Cost Energy Decisions for Buildings, Part I: Introduction to Life-Cycle Costing. Gaithersburg, MD: National Institute of Standards and Technology, April 1990.

Part I of this audiovisual series discuses life-cycle analysis as a tool to evaluate building design systems and equipment. It walks the audience through each step of the process. It explains the discount of cash flows, data sources, uncertainty, computer software, and the improvement of decisions. Part I of this series is also accompanied by a workbook: NISTIR 4309, *Least-Cost Energy Decisions for Buildings*.

4.1.1.21 Marshall, Harold E. Audiovisual series on Least-Cost Energy Decisions for Buildings, Part II: Uncertainty and Risk. Gaithersburg, MD: National Institute of Standards and Technology, April 1992.

Part II of this audiovisual series is an overview of uncertainty and risk. It discusses risk taking attitudes, risk neutral attitudes, and risk adverse attitudes. It explains methods to measure the probability that a project will successfully fulfill expectations. It also explains sensitivity analysis, probability, and simulations. It provides additional references and sources of data. Part II of this series is accompanied by a workbook: NISTIR 5178, *Least-Cost Energy Decisions for Buildings.*

4.1.1.22 Marshall, Harold E. Audiovisual series on Least-cost Energy Decisions for Buildings, part III: Choosing Economic Evaluation Methods. Gaithersburg, MD: National Institute of Standards and Technology.

Part III of this audiovisual series provides a method to select economic methods to evaluate types of investment decisions in buildings. It will allow the user to select from several economic methods: life-cycle cost, net savings, savings-to-investment ratio, adjusted internal rate of return, and payback methods. The types of investment decisions include the following topics: accept/reject, efficiency level, system design selection, optimal combination of projects, and priority ranking of projects. Part III also is accompanied by a workbook: *Least-Cost Energy Decisions for Buildings.*

4.1.1.23 Ramachandran, Ganapathy. The Economics of Fire Protection. New York, NY: Routledge, 1998.

This text provides methods for selecting a cost-effective fire protection strategy. It begins with a discussion of topics in fire protection economics: continuing with costs and benefits, cost-benefit analysis, trade-offs, decision analysis, utility theory, value-of-human life, fire insurance and analytical methods. These methods are directly relevant for those who are making mitigation decisions for fire protection. This text is mathematical in nature.

4.1.1.24 Ruegg, Rosalie T. and Harold Marshall. Building Economics: Theory and Practice. New York, NY: Van Nostrand Reinhold, 1990.

This text presents methods for economical decisions on the construction and renovation of buildings. It explains life-cycle cost, net benefits, net savings, benefit-to-cost ratio, savings-to-investment ratio, internal rate of return, overall rate of return, and payback. It describes the specific information that is needed to apply these techniques and how to treat uncertainty and risk. Topics include risk exposure, breakeven analysis, sensitivity analysis, decision analysis, and other techniques to treat risk and uncertainty.

4.1.1.25 Taylor, Craig, and Erik VanMarcke. Infrastructure Risk Management Processes: Natural, Accidental, and Deliberate hazards. ASCE Council on Disaster Risk Management Monograph No. 1. Reston, VA: American Society of Civil Engineers, 2006.

This monograph consists of eight papers that discuss the quantification of vulnerability decision analyses. It discusses hazards, systems evaluation, risk criteria, and systems management and includes the following titles:

- *PSHA* [Probabilistic Seismic Hazard Analysis] *Uncertainty Analysis: Applications to the CEUS and the Pacific NW*

- *The Regional Economic Cost of a Tsunami Wave generated by a Submarine Landslide off Palos Verdes, California*
- *The Emerging Role of Remote Sensing Technology in Emergency Management*
- *Context and Resiliency: Influences on Electric Utility Lifeline Performance*
- *Criteria for Acceptable Risk in the Netherlands*
- *Landslide Risk Assessment and Remediation*
- *A Preliminary Study of Geologic Hazards for the Columbia River Transportation Corridor*
- *Multihazard Mitigation Los Angeles Water System: A Historical Perspective*

4.1.2 Industry Standards

4.1.2.1 American National Standards Institute. "American National Standard for Occupational Health and Safety Management Systems." ANSI/AIHA Z10.

ANSI/AIHA Z10 is a voluntary consensus standard that provides "critical management systems requirements and guidelines for improvement of occupational health and safety." It affects health and safety as well as productivity, financial performance, and quality. Topics of ANSI/AIHA Z10 include participation, planning, implementation, evaluation, corrective action, and management review. This document is available through the American Industrial Hygiene Association: http://www.aiha.org.

4.1.2.2 ASTM International. ASTM Standards on Building Economics, 5th edition. West Conshohocken, PA: ASTM International, 2004.

This text includes all the standards associated with buildings and building projects. It includes the following ASTM standards: E 1185, E 1369, E 2204, E 917, E 964, E 1057, E 1074, E 1121, E 1699, E 1765, E 1804, E 1946, E 2013, E 2166, E 833, E 1557, E 2083, E 2103, E 2150, and E 2168. Each of these standards is described in section 5.1.2 of this guide. *ASTM Standards on Building Economic* is available for purchase from the ASTM International website: http://www.astm.org/.

4.1.2.3 ASTM International. "Standard Classification for Allowance, Contingency and Reserve Sums in Building Construction Estimating." E 2168. Annual Book of ASTM Standards: 2005. Vol. 04.12. West Conshohocken, PA: ASTM International.
URL: http://www.astm.org/

This standard is a classification to make monetary provisions for unexpected costs, risk, or other contingencies. Owners, developers, architects, and other professionals utilize this standard for cost budgeting.

4.1.2.4 ASTM International. "Standard Classification for Bridge Elements and Related Approach Work." E 2103. Annual Book of ASTM Standards: 2005. Vol. 04.12. West Conshohocken, PA: ASTM International.
URL: http://www.astm.org/

This is a standard for classifying bridge elements and labor. These are the major components that are common to most bridges regardless of material or design. Based on UNIFORMAT II (see reference 4.1.2.6 on page 83), it includes three levels of classification: Major Group Elements, Group Elements, and Individual Elements. Major Group Elements include items such as site work, substructure, and superstructure. Group Elements include items such as utility relocation, demolition, piers, and railings. Individual Elements include more specific items: sheeting, seal coat, dewatering, excavation, piles, and drilled shafts.

4.1.2.5 ASTM International, "Standard Classification for Building Construction Field Requirements, and Office Overhead and Profit," E 2083. Annual Book of ASTM Standards: 2005. Vol. 04.12. West Conshohocken, PA: ASTM International.
URL: http://www.astm.org/

This standard provides a classification for field requirements and office overhead and profit in construction estimating. These costs are common to all construction work and are an essential part of a complete cost estimate. This standard uses the UNIFORMAT II design from reference 4.1.2.6 (ASTM E 1557) and includes the first three levels of classification.

4.1.2.6 ASTM International. "Standard Classification for Building Elements and Related Site work-UNIFORMAT II." E 1557. Annual Book of ASTM Standards: 2005. Vol. 04.11. West Conshohocken, PA: ASTM International.
URL: http://www.astm.org/

ASTM E 1557 provides a standard for documentation and classification of building components and construction work. This format is used by owners, developers, planners, estimators, and other professionals and makes it easy to find and compare data. Well organized data is essential for any project. It includes four major classification elements: Major Group Elements, Group Elements, Individual Elements, and Sub-Elements. Each classification is more specific than the previous one. Examples of Major Group Elements include the substructure and the shell while Group Elements include foundation, basement construction and roofing. Individual Elements include items such as basement excavation, roof coverings, and roof openings. Sub-Elements are more specific items such as trenches, structure backfill, sheeting, and ramps. Also see MasterFormat in reference 4.1.2.30.

4.1.2.7 ASTM International. "Standard Classification for Life-Cycle Environmental Work Elements-Environmental Cost Element Structure." E 2150. Annual Book of ASTM Standards: 2005. Vol. 04.12. West Conshohocken, PA: ASTM International.
URL: http://www.astm.org/

This standard is based on the Interagency Environmental Cost Element Structure (ECES) and establishes a list of elements for life-cycle environmental work. It is a hierarchical standard format for costing so that information can be easily accessed. This standard is applicable to all environmental work, including waste management, environmental cleanup, decontamination, and other projects.

4.1.2.8 ASTM International. "Standard Guide for Developing a Cost-Effective Risk Mitigation Plan for New and Existing Constructed Facilities." ASTM E 2506. Annual Book of ASTM Standards: 2006. Vol. 04.12. West Conshohocken, PA: ASTM International.
URL: http://www.astm.org/

This guide describes a generic framework for developing a cost-effective risk mitigation plan for new and existing constructed facilities—buildings, industrial facilities, and other critical infrastructure. The guide provides owners and managers of constructed facilities, architects, engineers, constructors, other providers of professional services for constructed facilities, and researchers an approach for formulating and evaluating combinations of risk mitigation strategies. The guide insures that the combinations of mitigation strategies are formulated so that they can be rigorously analyzed with economic tools.

4.1.2.9 ASTM International. "Standard Guide for Selecting Economic Methods for Evaluating Investments in Buildings and Building Systems." ASTM E 1185. Annual Book of ASTM Standards: 2005. Vol. 04.11. West Conshohocken, PA: ASTM International.
URL: http://www.astm.org/

ASTM Standard E 1185 identifies the type of building decision and recommends ASTM practices and other supplements that should be used for a certain project. This ASTM standard is to be used at the initial stage of economic evaluation of a building project in order to identify the best assessment method.

4.1.2.10 ASTM International. "Standard Guide for Selecting Techniques for Treating Uncertainty and Risk in the Economic Evaluation of Buildings and Building Systems." E 1369. Annual Book of ASTM Standards: 2005. Vol. 04.11. West Conshohocken, PA: ASTM International.
URL: http://www.astm.org/

ASTM E 1369 provides methods for addressing uncertainty in the evaluation of buildings and building systems. It includes the uncertainty of the economic outcome and values used in the assessment of a building or building system. It discusses breakeven analysis, sensitivity analysis, risk-adjusted discounting, the mean-variance criterion and coefficient of variation, and decision analysis.

4.1.2.11 ASTM International. "Standard Guide for Summarizing the Economic Impacts of Building-Related Projects." E 2204. Annual Book of ASTM Standards: 2005. Vol. 04.12. West Conshohocken, PA: ASTM International.
URL: http://www.astm.org/

ASTM E 2204 details a generic format that uses 2 pages to summarize the economic evaluation of a building or building project. Sections of a typical summary include the significance of the project, key points, analysis strategy description, benefits and costs, key measures, and references.

4.1.2.12 ASTM International. "Standard Guide for the Estimation of Building Damageability in Earthquakes." E 2026. Annual Book of ASTM Standards: 2005. Vol. 04.12. West Conshohocken, PA: ASTM International.
URL: http://www.astm.org/

See reference 2.1.2 (page 8)

4.1.2.13 ASTM International. "Standard Practice for Applying Analytical Hierarchy Process (AHP) to Multi-attribute Decision Analysis of Investments Related to Buildings and building Systems." E 1765. Annual Book of ASTM Standards: 2005. Vol. 04.12. West Conshohocken, PA: ASTM International.
URL: http://www.astm.org/

The analytical hierarchy process (AHP) described in this standard is a multi-attribute decision analysis (MADA) method. Using pair-wise comparisons it allows the evaluation of monetary as well as non-monetary benefits and costs. The standard also describes a software package available to assist in the analytical hierarchy process.

4.1.2.14 ASTM International. "Standard Practice for Constructing FAST Diagrams and Performing Function Analysis during Value Analysis Study." E 2013. Annual Book of ASTM Standards: 2005. Vol. 04.12. West Conshohocken, PA: ASTM International.
URL: http://www.astm.org/

This standard provides a structure to analyze a building project. Unnecessary costs are identified by determining low preference/high cost elements and high preference/low cost elements. It also establishes a communication format that is understood by owners, users, and stakeholders.

4.1.2.15 ASTM International. "Standard Practice for Measuring Benefit-to-Cost and Savings-to-Investment Ratios for Buildings and Building Systems." E 964. Annual Book of ASTM Standards: 2005. Vol. 04.11. West Conshohocken, PA: ASTM International.
URL: http://www.astm.org/

ASTM standard E 964 details the benefit-to-cost ratio (BCR) and savings-to-investment ratio (SIR). These two methods are used to evaluate buildings and building systems in relation to benefits and costs. There are several other methods for assessing buildings: internal rate of return, net benefits, payback, multi-attribute decision analysis, and risk analysis. These measures are detailed in other ASTM standards.

4.1.2.16 ASTM International. "Standard Practice for Measuring Cost Risk of Buildings and Building Systems." E 1946. Annual Book of ASTM Standards: 2005. Vol. 04.12. West Conshohocken, PA: ASTM International.
URL: http://www.astm.org/

ASTM E 1946 establishes a standard for measuring cost risk for buildings and building systems using the Monte Carlo simulation technique. This method is a standard practice for cost information, but requires computer software to conduct the analysis. It provides probabilities for construction costs to be above or below the estimated value; the price range of construction costs; and identifies the elements with the greatest impact on construction costs.

4.1.2.17 ASTM International. "Standard Practice for Measuring Internal Rate of Return and Adjusted Internal Rate of Return for Investments in Buildings and Building Systems." E 1057. Annual Book of ASTM Standards: 2005. Vol. 04.11. West Conshohocken, PA: ASTM International.
URL: http://www.astm.org/

This standard discusses the internal rate of return and adjusted internal rate of return, which measures the economic return on an investment in a building or building system. This could be compared to a discount rate or the real interest rate on an investment. There are several other methods for assessing building investments: benefit-to-cost ratio, life-cycle cost, net benefits, payback analysis, multi-attribute decision analysis, and risk analysis.

4.1.2.18 ASTM International. "Standard Practice for Measuring Life-Cycle Costs of Buildings and Building Systems." E 917. Annual Book of ASTM Standards: 2005. Vol. 04.11. West Conshohocken, PA: ASTM International.
URL: http://www.astm.org/

ASTM Standard E 917: "Standard Practice for Measuring Life-Cycle Costs of Buildings and Building Systems" details the life cycle cost method as it relates to buildings and their systems. Several other methods for assessing buildings include benefit-to-cost ratio, internal rate of return, net benefits, payback, multi-attribute decision analysis, and risk analysis. These measures are detailed in other ASTM standards.

4.1.2.19 ASTM International. "Standard Practice for Measuring Net Benefits and Net Savings for Investments in Buildings and Building Systems." E 1074. Annual Book of ASTM Standards: 2005. Vol. 04.11. West Conshohocken, PA: ASTM International.
URL: http://www.astm.org/

ASTM E 1074 details the process of measuring net benefits (net savings), which is the discounted difference between the benefits (savings) of a project and the costs. This method measures the cost effectiveness of a building or building system and can be used to compare building designs and systems or to evaluate a single design. There are several other methods for assessing building investments: benefit-to-cost ratio, life-cycle cost, internal rate of return, payback analysis, multi-attribute decision analysis, and risk analysis.

4.1.2.20 ASTM International. "Standard Practice for Measuring Payback for Investments in Buildings and Building Systems." E 1121. Annual Book of ASTM Standards: 2005. Vol. 04.11. West Conshohocken, PA: ASTM International.
URL: http://www.astm.org/

ASTM E 1121 is the standard practice for measuring payback, which is the period of time that it takes for the benefits (i.e., savings) incurred from a project to equal the cost of a project. This can be used to identify optimal building designs and systems. There are several other methods for assessing building investments: benefit-to-cost ratio, life-cycle cost, internal rate of return, net benefits, multi-attribute decision analysis, and risk analysis.

4.1.2.21 ASTM International. "Standard Practice for Organizing and Managing Building Data." E 2166. Annual Book of ASTM Standards: 2005. Vol. 04.12. West Conshohocken, PA: ASTM International.
URL: http://www.astm.org/

This practice provides a standard method to organize building information, which includes text, numeric, and graphical data. It can be applied to information for a single building or multiple buildings and is based on UNIFORMAT II (ASTM E 1557 in reference 4.1.2.6). Since this format makes it easy to find and compare data many software programs use it.

4.1.2.22 ASTM International. "Standard Practice for Performing and Reporting Cost Analysis during the Design Phase of a Project." E 1804. Annual Book of ASTM Standards: 2005. Vol. 04.12. West Conshohocken, PA: ASTM International.
URL: http://www.astm.org/

ASTM E 1804 is an industry standard for providing cost analysis in the design phase of a project. It is intended to increase communication between the design professional, owner, and the consultants.

4.1.2.23 ASTM International. "Standard Practice for Performing Value Analysis (VA) of Buildings and Building Systems." E 1699. Annual Book of ASTM Standards: 2005. Vol. 04.11. West Conshohocken, PA: ASTM International.
URL: http://www.astm.org/

This industry standard details a procedure for satisfying the criteria of a project. These requirements are then linked to cost and benefits. There are three periods of activity described in this standard: preparation effort, workshop effort, and post workshop effort.

4.1.2.24 ASTM International. "Standard Terminology of Building Economics." E 833. Annual Book of ASTM Standards: 2005. Vol. 04.11. West Conshohocken, PA: ASTM International.
URL: http://www.astm.org/

ASTM E 833 provides a standard set of over 100 terms and definitions for building economics.

4.1.2.25 American Society of Civil Engineers. Flood Resistant Design and Construction. SEI/ASCE 24-05. Reston, VA: American Society of Civil Engineers, 2003.

See reference 3.1.7 (page 35)

4.1.2.26 American Society of Civil Engineers. Minimum Design Loads for Buildings and Other Structures. SEI/ASCE 7-05. Reston, VA: American Society of Civil Engineers, 2003.

See reference 3.1.8 (page 35)

4.1.2.27 American Society of Civil Engineers. Seismic Evaluation of Existing Buildings. SEI/ASCE 31-03. Reston, VA: American Society of Civil Engineers, 2003.

See reference 2.1.3 (page 8)

4.1.2.28 American Society of Professional Estimators. Standard Estimating Practice.
URL: http://www.aspenational.com

This text provides a reference manual arranged in CSI format for estimating construction practices. It is available for purchase on the American Society of Professional Estimators' website listed above. CSI provides education and fellowship for construction estimators and has two certification programs: ASPE Certification and CPE Recognition.

4.1.2.29 British Standards Institute.
URL: http://www.bsi-global.com

The British Standards Institute is a national standards body of the UK. It develops standards and solutions for business and society. They develop standards for building and construction, the environment, fire codes, food, health and safety, healthcare, manufacturing, risk, and security. They also provide a number of education courses.

4.1.2.30 Construction Specifications Institute.
URL: http://www.csinet.org

The Construction Specifications Institute (CSI) conducts and publishes research to improve the built environment. It develops standards such as the National CAD Standard and MasterFormat, which is used for detailing and developing construction cost data. Also see UNIFORMAT in reference 4.1.2.6.

4.1.2.31 International Organization of Standardization.
URL: http://www.iso.org

The International Organization of Standardization (ISO) is a world standards organization. It works with national and regional standards organization to develop international standards that facilitate trade, improve quality, improve security, protect the environment, protect the consumer, and promotes economic growth. ISO has developed standards for manufacturing, engineering, telecommunications, information technology, food, construction, military engineering, and a number of other fields.

4.1.2.32 National Fire Protection Association.

See reference 2.3.2.9 (page 28)

4.1.2.33 National Fire Protection Association. Building and Construction Safety Code. NFPA 5000. Quincy, MA: National Fire Protection Association, 2003.
URL: http://www.nfpa.org

This publication specifies building systems for health, safety, comfort, and convenience for every aspect of design and construction of buildings. There are a total of 55 chapters that address construction type and height requirements, fire-resistive materials, interior finish, means of egress, accessibility, building rehabilitation, and a number of other topics. This text is available for purchase from the National Fire Protection Association at the web address cited above.

4.1.2.34 United States General Services Administration. Facilities Standards for the Public Buildings Service. P100. March 2005.
URL: http://www.gsa.gov/Portal/gsa/ep/home.do?tabId=1

See reference 3.1.80 (page 54)

4.1.3 Software for Implementing Industry Standards

4.1.3.1 National Institute of Standards and Technology (NIST): Building and Fire Research Laboratory.
URL: http://www.bfrl.nist.gov

See reference 2.2.10 (page 18)

4.1.3.2 National Institute of Standards and Technology (NIST): Building and Fire Research Laboratory. Building for Environmental and Economic Sustainability.
URL: http://www.bfrl.nist.gov/info/software.html

The Building for Environmental and Economic Sustainability (BEES) software provides a way to select cost-effective and environmentally preferable building products. Developed by NIST's Building and Fire Research Laboratory, BEES includes actual environmental and economic performance data for almost 200 building products. It uses a life-cycle assessment approach to measure the environmental performance of each product. Every stage of the products life is evaluated: raw material, manufacture, transportation, installation, use, and disposal and recycling. BEES is supported in part by the U.S. EPA Environmentally Preferable Purchasing (EPP) Program.

4.1.3.3 National Institute of Standards and Technology (NIST): Building and Fire Research Laboratory. Building Life Cycle Cost.
URL: http://www.bfrl.nist.gov/info/software.html

Building Life-Cycle Cost (BLCC) is a software program that assists users in making building and building project decisions by calculating net savings, savings-to-investment ratio, adjusted internal rate of return, life-cycle cost, and years to payback. Handbook 135 (see reference 4.1.1.18 on page 80) as well as ASTM Standards on Building Economics (see reference 4.1.2.2 on page 82) describes and details the life-cycle cost method. BLCC is available for download free to the public.

4.1.3.4 National Institute of Standards and Technology (NIST): Building and Fire Research Laboratory. Cost Effectiveness Software Tool.
URL: http://www.bfrl.nist.gov/info/software.html

The Cost-Effectiveness Software Tool assists in making consistent comparisons of risk mitigation strategies. It allows the user to define hazard scenarios, identify consequences, and compare combinations of strategies to mitigate outcomes. The software is based on industry standard ASTM E 917 and compares life-cycle cost, adjusted internal rate of return, present value net savings, and savings to investment ratio. In order to calculate and compare each mitigation strategy the user is required to enter data for costs, events, discount rate, escalation rates, and cost frequencies. These terms are discussed and explained within the software as well as in the ASTM standard E 917. This user friendly software is available to download for free at the NIST website.

4.1.3.5 National Institute of Standards and Technology (NIST): Building and Fire Research Laboratory. Energy Escalation Rate Calculator.
URL: http://www.bfrl.nist.gov/oae/software.html

The Office of Applied Economics provides an Energy Escalation Rate Calculator (EERC) that computes an average annual escalation rate for fuel prices using forecast information from the DOE. The user enters the project location, fuel usage, industry sector, and the dates of the performance period. Weighted by the proportions of fuel type, the software calculates the average rate of fuel price escalation over the duration of the performance period. This software compliments the Building Life Cycle Cost software (see reference 4.1.3.3 on page 90).

4.2 Economic Modeling Resources

4.2.1 ASTM International. ASTM Standards on Building Economics, 5th edition. West Conshohocken, PA: ASTM International, 2004.

See reference 4.1.2.2 (page 82)

4.2.2 ASTM International. "Standard Guide for Selecting Economic Methods for Evaluating Investments in Buildings and Building Systems." ASTM E 1185. Annual Book of ASTM Standards: 2005. Vol. 04.11. West Conshohocken, PA: ASTM International.
URL: http://www.astm.org/

See reference 4.1.2.9 (page 84)

4.2.3 ASTM International. "Standard Guide for Selecting Techniques for Treating Uncertainty and Risk in the Economic Evaluation of Buildings and Building Systems." E 1369. ASTM Annual Book of ASTM Standards: 2005. Vol. 04.11. West Conshohocken, PA: ASTM International.
URL: http://www.astm.org/

See reference 4.1.2.10 (page 84)

4.2.4 ASTM International. "Standard Guide for Summarizing the Economic Impacts of Building-Related Projects." E 2204. Annual Book of ASTM Standards: 2005. Vol. 04.12. West Conshohocken, PA: ASTM International.
URL: http://www.astm.org/

See reference 4.1.2.11 (page 85)

4.2.5 ASTM International. "Standard Practice for Applying Analytical Hierarchy Process (AHP) to Multi-attribute Decision Analysis of Investments Related to Buildings and Building Systems." E 1765. Annual Book of ASTM Standards: 2005. Vol. 04.12. West Conshohocken, PA: ASTM International.
URL: http://www.astm.org/

See reference 4.1.2.13 (page 85)

4.2.6 ASTM International. "Standard Practice for Measuring Benefit-to-Cost and Savings-to-Investment Ratios for Buildings and Building Systems." E 964. Annual Book of ASTM Standards: 2005. Vol. 04.11. West Conshohocken, PA: ASTM International.
URL: http://www.astm.org/

See reference 4.1.2.15 (page 86)

4.2.7 ASTM International. "Standard Practice for Measuring Cost Risk of Buildings and Building Systems." E 1946. Annual Book of ASTM Standards: 2005. Vol. 02.12. West Conshohocken, PA: ASTM International.
URL: http://www.astm.org/

See reference 4.1.2.16 (page 86)

4.2.8 ASTM International. "Standard Practice for Measuring Internal Rate of Return and Adjusted Internal Rate of Return for Investments in Buildings and Building Systems." E 1057. Annual Book of ASTM Standards: 2005. Vol. 02.11. West Conshohocken, PA: ASTM International.
URL: http://www.astm.org/

See reference 4.1.2.17 (page 86)

4.2.9 ASTM International. "Standard Practice for Measuring Life-Cycle Costs of Buildings and Building Systems." E 917. Annual Book of ASTM Standards: 2005. Vol. 04.11. West Conshohocken, PA: ASTM International.
URL: http://www.astm.org/

See reference 4.1.2.18 (page 86)

4.2.10 ASTM International. "Standard Practice for Measuring Net Benefits for Investments in Buildings and Building Systems." E 1074. Annual Book of ASTM Standards: 2005. Vol. 04.11. West Conshohocken, PA: ASTM International.
URL: http://www.astm.org/

See reference 4.1.2.19 (page 87)

4.2.11 ASTM International. "Standard Practice for Measuring Payback for Investments in Buildings and Building Systems." E 1121. Annual Book of ASTM Standards: 2005. Vol. 04.11. West Conshohocken, PA: ASTM International.
URL: http://www.astm.org/

See reference 4.1.2.20 (page 87)

4.2.12 ASTM International. "Standard Practice for Performing Value Analysis (VA) of Buildings and Building Systems." E 1699. Annual Book of ASTM Standards: 2005. Vol. 04.11. West Conshohocken, PA: ASTM International.
URL: http://www.astm.org/

See reference 4.1.2.23 (page 88)

4.2.13 Federal Reserve Bank.
URL: http://www.federalreserve.gov/

The central bank of the United States, the Federal Reserve Bank, is responsible for conducting monetary policy, regulating bank institutions, and providing some additional financial services. A significant amount of economic data that is commonly used by economists for research is available free to the public from the Federal Reserve's website: http://www.federalreserve.gov/datadownload/. It includes a large range of information: industry debt, industry production, flow of funds, and selected interest rates. Two economic databases can be accessed through the Federal Reserve Bank of St Louis: the Federal Reserve Economic Database (FRED) and the ArchivaL Federal Reserve Economic Database (ALFRED). Both include banking data, interest rates, consumer price indexes, employment data, population data, exchange rates, transactions, producer price indexes, and business data. The data is free to the public at http://www.stlouisfed.org/.

4.2.14 Fuller, Sieglinde K. Energy Price Indices and Discount Factors for Life-Cycle Cost Analysis: Annual Supplement to NIST Handbook 135 and NBS Special Publication 709, NISTIR 85-3273-21. Gaithersburg, MD: National Institute of Standards and Technology, April 2006.
URL: http://www1.eere.energy.gov/femp/pdfs/ashb06.pdf

The *Energy Price Indices and Discount Factors for Life-Cycle Cost Analysis* is an annual publication. It supports the life-cycle cost methodology described in ASTM Standard E-917 through the updating of the energy price projections and discount factors, which is needed for calculating life-cycle costs.

4.2.15 Fuller, Sieglinde K., and Stephen R. Petersen. Life-Cycle Costing Manual for the Federal Energy Management Program. Hand Book 135, 1995 Edition. Gaithersburg, MD: National Institute of Standards and Technology, 1995.
ULR: http://fire.nist.gov/bfrlpubs/build96/PDF/b96121.pdf

See reference 4.1.1.18 (page 80)

4.2.16 Mackin, T.J., LTC Darrall Henderson, and J.W. Jones. A Method For Allocating Financial Resources to Combat Terrorism: Optimizing the Reduction of Consequences. 71st MORS Symposium, Working Group 16, June 13, 2003. Alexandria, VA: Military Operations Research Society, October 2003.
URL: http://handle.dtic.mil/100.2/ADA418274

This paper presents a mathematical expression for resource allocation to decrease vulnerability to terrorist attacks. It allows decision-makers to optimize the allocation of resources for risk mitigation against terrorist attacks. This article starts by establishing a basic approach to rank terrorist threats. It is available from the Homeland Security Digital Library discussed in reference 3.1.75 (https://www.hsdl.org)

4.2.17 Marshall, Harold E. Audiovisual series on Least-Cost Energy Decisions for Buildings, Part I: Introduction to Life-Cycle Costing. Gaithersburg, MD: National Institute of Standards and Technology, April 1990.

See reference 4.1.1.20 (page 80)

4.2.18 Marshall, Harold E. Audiovisual series on Least-Cost Energy Decisions for Buildings, Part II: Uncertainty and Risk. Gaithersburg, MD: National Institute of Standards and Technology, April 1992.

See reference 4.1.1.21 (page 80)

4.2.19 Marshall, Harold E. Audiovisual series on Least-cost Energy Decisions for Buildings, part III: Choosing Economic Evaluation Methods. Gaithersburg, MD: National Institute of Standards and Technology.

See reference 4.1.1.22 (page 81)

**4.2.20 NIST. Energy Escalation Rate Calculator.
URL: http://www.bfrl.nist.gov/oae/software.html**

See source in section 4.1.3.5.

4.2.21 Rogers, Mark R. Handbook of Key Economic Indicators. Chicago, IL: Irwin Professional Publishing, 1994.

The *Handbook of Key Economic Indicators* explains, defines, and illustrates economic data that is used to assess the economy. It discusses issues such as the employment report, personal income report, consumer price index, producer price index, and gross domestic product. The data in this text is used to evaluate the past, present, and future state of the economy. It has a significant influence on capital returns (i.e., discount rates).

4.2.22 Ruegg, Rosalie T. and Harold Marshall. Building Economics: Theory and Practice. New York, NY: Van Nostrand Reinhold, 1990.

See reference 4.1.1.24 (page 81)

**4.2.23 United States Department of Commerce: Bureau of Economic Analysis.
URL: http://www.bea.gov/**

Promoting a better understanding of the U.S. economy, the Bureau of Economic Analysis (BEA) provides economic accounts data that is used by decision-makers, researchers, and the American public. Private and public sector individuals closely watch some of the data provided by BEA. It is used by the White House, Congress, Federal Reserve, and Wall Street as an indicator of the economy. The Department of Commerce recognized the GDP as the bureau's greatest achievement and as one of the three most influential measures that affect the U.S. financial market. The BEA has four economic accounts programs: National Accounts, which provides insight into production, consumption, investment, and saving; Regional Accounts, which generates data on economic activity by region, state, and county; Industry Accounts, which provides industry input and output tables; and International Accounts, which provides balance of payments data.

**4.2.24 United States Department of Labor: Bureau of Labor Statistics.
URL: http://www.bls.gov/**

The Bureau of Labor Statistics (BLS) provides detailed historic and current statistics on national inflation, the consumer price index, wages, employment, productivity, safety and health, demographics, and more. It is an important agency that collects economic and labor data for the federal government. BLS data is commonly used by economists and can be easily downloaded for statistical analysis.

4.3 Analysis Strategies for Treating Uncertainty

4.3.1 ASTM International. "Standard Guide for Developing a Cost-Effective Risk Mitigation Plan for New and Existing Constructed Facilities." ASTM E 2506. Annual Book of ASTM Standards: 2006. Vol. 04.12. West Conshohocken, PA: ASTM International.
URL: http://www.astm.org/

See reference 4.1.2.8 (page 84)

4.3.2 ASTM International. "Standard Guide for Selecting Economic Methods for Evaluating Investments in Buildings and Building Systems." ASTM E 1185. Annual Book of ASTM Standards: 2005. Vol. 04.11. West Conshohocken, PA: ASTM International.
URL: http://www.astm.org/

See reference 4.1.2.9 (page 84)

4.3.3 ASTM International. "Standard Guide for Selecting Techniques for Treating Uncertainty and Risk in the Economic Evaluation of Buildings and Building Systems." E 1369. Annual Book of ASTM Standards: 2005. Vol. 04.11. West Conshohocken, PA: ASTM International.
URL: http://www.astm.org/

See reference 4.1.2.10 (page 84)

4.3.4 ASTM International. "Standard Practice for Measuring Cost Risk of Buildings and Building Systems." E 1946. Annual Book of ASTM Standards: 2005. Vol. 04.12. West Conshohocken, PA: ASTM International.
URL: http://www.astm.org/

See reference 4.1.2.16 (page 86)

4.3.5 ASTM International. "Standard Practice for Measuring Life-Cycle Costs of Buildings and Building Systems." E 917. Annual Book of ASTM Standards: 2005. Vol. 04.11. West Conshohocken, PA: ASTM International.
URL: http://www.astm.org/

See reference 4.1.2.18 (page 86)

4.3.6 Grossi, Patricia, Howard Kunreuther, and Chandu C. Patel. Catastrophe Modeling: A New Approach to Managing Risk. New York, NY: Springer Science and Business Media, 2005.

See reference 3.1.39 (page 43)

4.3.7 Marshall, Harold E. Techniques for Treating Uncertainty and Risk in the Economic Evaluation of Building Investments. NIST Special Publication 757. Gaithersburg, MD: National Institute of Standards and Technology, 1988.

This publication details uncertainty in input values and techniques that measure the risk that a project will have an unfavorable economic outcome. It discusses risk exposure and attitude along with breakeven analysis, sensitivity analysis, risk-adjusted discount rate, and decision analysis.

4.3.8 Marshall, Harold E., Robert E. Chapman, and Chi J. Leng. "Risk Mitigation Plan for Optimizing Protection of Constructed Facilities." Cost Engineering. Vol. 46, No. 8 (August 2004) 26.

See reference 3.1.47 (page 46)

4.3.9 Stewart, Mark G., and Robert E. Melchers. Probabilistic Risk Assessment of Engineering Systems. London: Chapman and Hall, 1997.

This text is a multi-disciplinary guide that enables the layperson to gain understanding of risk analysis procedures. It applies to making decisions about the viability of an engineered system. To be a viable system it must be economical, fulfill its specified purpose, and perform at a specified level of safety. The term "system" can include "nuclear power plants, railway systems, bridges, petrochemical installations and many others." This text guides the reader in evaluating a system, the probability that it fulfills its necessary requirements and the consequences if it does not fulfill those requirements.

5 Summary and Recommendations for Further Research

5.1 Summary

This report supports the development of a cost-effective risk mitigation plan that involves (1) assessing the risks associated with natural and man-made hazards; (2) formulating combinations of mitigation strategies for constructed facilities exposed to those hazards; and (3) using economic tools to identify the most cost-effective combination of strategies. Furthermore, it asserts that developing a risk mitigation plan requires both guidance and data. Guidance is needed to help owners and managers to assess the risks facing their facility. Data about the frequency and consequences of natural and man-made hazards are needed when assessing the risks that a particular facility faces from these hazards. Estimates of the costs of protection are needed to ensure that safeguarding personnel and physical assets and satisfying financial constraints are kept in balance. Finally, guidance on the use of economic evaluation methods is needed to ensure that the correct method, or combination of methods, is used.

The need for this report stems from the fact that although there is a great deal of high-quality information available on risk assessment and risk management, natural and man-made hazards, and economic tools, there is no central source of data and tools to which the owners and managers of constructed facilities can turn for help in developing a cost-effective risk mitigation plan. This report fulfills that need by providing an annotated bibliography of printed and electronic resources that serves as a central source of data and tools to help the owners and managers of constructed facilities develop a cost-effective risk mitigation plan.

5.2 Recommendations for Further Research

Given that new security-related materials are being published on a regular basis, it is recommended that this report be revised and reissued periodically. Periodic revision and reissue would enable new references to be incorporated and web links to be added, revised, or deleted, as needed. Periodic revision would ensure that owners and managers of constructed facilities would have access to a body of information that is pre-packaged in a format that well serves their need for making cost-effective responses to multiple hazards and timely in its coverage of this rapidly emerging field of vital importance to homeland security.

Appendix A: Overview of the Three-Step Protocol for Developing a Cost-Effective Risk Mitigation Plan

Protecting constructed facilities from damages from natural and man-made hazards in a cost-effective manner is a challenging task. The three-step protocol was designed by Chapman and Leng[6] to address this challenge. A three-step protocol provides owners and managers of constructed facilities, architects, engineers, constructors, other providers of professional services for constructed facilities, and researchers an approach for formulating and evaluating combinations of risk mitigation strategies. The three steps that decision-makers follow in the protocol are to (1) assess the likelihood that their facility and its contents will be damaged from natural and man-made hazards; (2) identify engineering, management, and financial strategies for abating the risk of damages; and (3) use standardized economic evaluation methods to select the most cost-effective combination of risk mitigation strategies to protect their facility.

The end product of the three-step protocol is a cost-effective risk mitigation plan for either new or existing constructed facilities. The purpose of the risk mitigation plan is to provide the most cost-effective reduction in personal injuries, financial losses, and damages to new and existing constructed facilities. Thus, the risk mitigation plan incorporates perspectives from multiple stakeholders—owners and managers, occupants and users, and other affected parties—in addressing natural and man-made hazards.

The three-step protocol is most effective when used in the project initiation and planning phases of the project delivery process. Consideration of alternative combinations of risk mitigation strategies early in the project delivery process allows both greater flexibility in addressing specific hazards and lower costs associated with their implementation.

The three-step protocol insures that the combinations of mitigation strategies are formulated so that they can be rigorously analyzed with economic tools. Economic tools include evaluation methods, standards that support and guide the application of those methods, and software for implementing the evaluation methods.

The organization of the three-step protocol is summarized in Table A.1. The table lists each key component and its associated subcomponents. Each key component in Table A.1 is a section in this chapter. The subcomponents map into this chapter's subsections. Table A.1 provides a snapshot of the three-step protocol that is useful in visualizing how the key components and subcomponents build on and reinforce each other. Throughout this appendix, references are given to other chapters to highlight links to the guidance documents, software, and data needed to implement that key component/subcomponent of the protocol.

---

[6] Chapman, Robert E. and Leng, Chi J. *Cost-Effective Responses to Terrorist Risks in Constructed Facilities*, NISTIR 7073 (Gaithersburg, MD: National Institute of Standards and Technology, March 2004).

Table A.1   Organization of the Three-Step Protocol

Key Component	Subcomponent
Perform Risk Assessment	Establish Risk Mitigation Objectives and Constraints
	Conduct Assessment and Document Findings
Specify Combinations of Risk Mitigation Strategies for Evaluation	Review Alternative Risk Mitigation Strategies
	Select Candidate Combinations of Risk Mitigation Strategies
	Develop Cost/Loss Estimates and Sequence of Cash Flows for Each Candidate Combination
Perform Economic Evaluation	Select Appropriate Economic Method(s) for Evaluating the Candidate Combinations of Risk Mitigation Strategies
	Compute Measures of Economic Performance for Each Candidate Combination
	Recompute Measures of Economic Performance Taking into Consideration Uncertainty
	Analyze Results and Identify the Most Cost-Effective Combination of Risk Mitigation Strategies
	Prepare Report with Documentation Supporting Recommended Risk Mitigation Plan

Terminology Used in the Three-Step Protocol

In order to gain a more complete understanding of the three-step protocol, it is useful to define several key terms. The terms are arranged in a logical order. Basic concepts are introduced first; they are used to build a foundation for the more complex terms that follow.

Constructed Facility—Any building, industrial facility, or component of the physical infrastructure. Examples of constructed facilities are: commercial and government office buildings, health care facilities, educational facilities, refineries, chemical manufacturing facilities, power plants, bridges, and tunnels.

Asset—Any constructed facility, its contents (physical systems, information, and personnel), or activities that have positive value to an owner or society as a whole.

Hazard—Any phenomenon capable of causing damage to an asset. Hazards fall into two basic types: natural and man-made.

Event—A means for classifying hazards. Associated with events are outcomes, vulnerabilities, and consequences.

Outcome—The results of an event. Outcomes are characterized by their severity and their occurrence probabilities. An event includes the full range of outcomes from no damage to extreme damage. For a given event, the sum of all outcome probabilities equals 1.0. Thus the no damage outcome would usually have a very high probability, corresponding to maintenance of the status quo, whereas an extreme damage outcome (e.g., a Category 5 hurricane making landfall within a specified radius of an asset) would have a very low probability.

Vulnerability—Any weakness in an asset's design, implementation, or operation that can result in damage to the asset. Such weaknesses can occur in facility characteristics, equipment properties, personnel behavior, locations of people and equipment, or operational and personnel practices.

Consequence—The immediate, short-, and long-term effects of an event-outcome combination. The consequence of an event-outcome combination on an asset is generally estimated as the amount of loss or damage that can be expected from that combination. Loss may be monetary, but may also include less tangible and therefore less easily quantified effects. Some examples of relevant consequences include: public or asset personnel (e.g., occupants, users, or third parties) fatalities or injuries, property damage or loss, disruption of public or private operations, environmental damage, and loss of critical data.

Risk—The product of the probability and consequences of an event-outcome combination summed across all event-outcome combinations. This definition of risk refers to multiple hazards. Risk can also be expressed with regard to a single hazard or with regard to a specific event-outcome combination (e.g., a Category 5 hurricane making landfall within a specified radius of an asset).

Risk Mitigation Strategy—Preplanned and coordinated actions or system features that are designed to: reduce the damage caused by an event-outcome combination and/or reduce the occurrence probability of that event-outcome combination, support and complement first responders, facilitate field investigation and crisis management response, and facilitate recovery. Risk mitigation strategies may be classified under three broad headings: engineering, managerial, and financial.

Economic Evaluation Methods—A set of economic analysis techniques that consider all relevant costs associated with a project investment during its study period, comprising such techniques as life-cycle cost, savings-to-investment ratio, adjusted internal rate of return, and net savings.

Cost-Accounting Framework (Mitigation Costs, Event-Related Losses, Cost Item)—A methodology for tracking how costs affect stakeholders (e.g., owners/managers, users/customers, third parties). The cost-accounting framework promotes a detailed, consistent breakdown of life-cycle costs. It is a means for facilitating the decision-making process by identifying unambiguously who bears which costs, how costs are allocated among several widely-accepted budget categories, and how costs are allocated among the three types of risk mitigation strategies. The cost-accounting framework makes a clear distinction between mitigation costs (i.e., the cost to implement and maintain a given combination of risk mitigation strategies) and event-related losses (i.e., for a given combination of risk mitigation strategies, the expected value of the product of the probability and consequences of an event-outcome combination summed over all event-outcome combinations). Event-related losses are a measure of risk. A cost item is a name/description assigned to a mitigation cost/event-related loss associated with a specific combination of risk mitigation strategies.

Multi-Hazard Approach—A comprehensive approach that stresses evaluating the full range of natural and man-made hazards affecting an asset as a group. The multi-hazard approach assumes that costs for protection against multiple hazards can be shared among the hazards protected against, thereby reducing the cost of protection for any single hazard.

Alternative—A combination of risk mitigation strategies. The alternative with the lowest first cost is frequently referred to as the base case against which other alternatives are compared.

Uncertainty—Lack of certain, deterministic, values for the variable inputs used in an economic evaluation of a competing set of alternative combinations of risk mitigation strategies. Treatment of uncertainty begins with the establishment of a baseline analysis, in which all inputs are set at their "best-guess" values. A sensitivity analysis is then conducted whereby one or more key inputs are systematically varied about their base line values. Monte Carlo simulation is then used to develop probabilistic levels of significance for key measures of economic performance (e.g., life-cycle costs).

Financial Risk—The probability of investing in a project whose economic performance is different from what is desired of expected. Financial risk is a byproduct of the Monte Carlo simulation.

Unquantified Effects—Factors that are an integral part of the decision process but are difficult or impossible to express as dollar amounts.

Cost-Effective Risk Mitigation Plan—That combination of risk mitigation strategies (i.e., alternative) that minimizes life-cycle costs.

A.1   Perform Risk Assessment

The first step in creating a cost-effective risk mitigation plan is a risk assessment for the facility or group of facilities to be protected. This step includes specification of the decision-maker's objectives (e.g., understand the nature of the risk to optimize the use of mitigation strategies), the facilities to be protected, the natural and man-made hazards to be considered, the composition of the risk assessment team, and documentation procedures. The risk assessment involves data collection to establish the likelihood of natural and man-made hazards as well as the on-site collection and documentation of facility vulnerabilities to those hazards. Estimates of the value of the facility's assets and the consequences of an event occurring are also produced as part of the risk assessment.

A.1.1 Establish Risk Mitigation Objectives and Constraints

Specify the decision-maker's objectives. This is crucial in defining the problem and determining the suitability of the economic evaluation method(s).

Identify the constructed facility or set of facilities to be evaluated. Identify the types of hazards to be evaluated.

Specify the design or system objective that is to be accomplished. Identify any constraints that limit the available options to be considered.

A.1.2 Conduct Assessment and Document Findings

Form an assessment team composed of individuals familiar with the type of facility or set of facilities to be evaluated, individuals familiar with assessment tools and techniques, and individuals who have breadth and depth of experience and understand other disciplines and system interdependencies. Refer to the risk assessment guidance documents and software tools summarized in Chapter 2 to gain assessment insights on specific hazards or classes of hazards. Supplement your data sources with those described in Section 2.3 to compile information on the likelihood and severity of specific hazards or classes of hazards.

Use information from the documents and software summarized in Chapter 2 to produce an assessment plan. Provide the assessment team with the tools, such as laptop computers and electronic forms/data collection sheets, needed to implement the assessment plan.

Make assignments and deploy the assessment team. Collect and compile information on the facility's assets and develop estimates of the value of the facility's assets. Collect and compile information on specific hazard types, their likelihood, facility vulnerabilities, and consequences.

Use an agreed upon format, such as ASTM Standard Classifications E 1557[7] or E 2103[8] or ASTM Standard Practice E 2166,[9] to create a compiled set of information collected

---

[7] ASTM International. "Standard Classification for Building Elements and Related Sitework—UNIFORMAT II," E 1557, *Annual Book of ASTM Standards: 2005*. Vol. 04.11. West Conshohocken, PA: ASTM International.

[8] ASTM International. "Standard Classification for Bridge Elements and Related Approach Work," E 2103, *Annual Book of ASTM Standards: 2005*. Vol. 04.12. West Conshohocken, PA: ASTM International.

[9] ASTM International. "Standard Practice for Organizing and Managing Building Data," E 2166, *Annual Book of ASTM Standards: 2005*. Vol. 04.12. West Conshohocken, PA: ASTM International.

from the assessment team that documents the findings of the risk assessment. Transmit the compiled set of information to a central repository to ensure that access to sensitive information can be limited to those with a legitimate need to know.

A.2   Specify Combinations of Risk Mitigation Strategies for Evaluation

The second step of the protocol focuses on identification of risk mitigation strategies. This step uses information from the risk assessment (e.g., estimates of the value of the facility's assets and the consequences of an event occurring) to identify engineering, management and financial strategies to mitigate those consequences. The costs of implementing the alternative risk mitigation strategies and the associated reductions in consequences are also produced as part of this step in the protocol.

A.2.1   Review Alternative Risk Mitigation Strategies

This section describes three risk mitigation strategies—engineering, management, and financial. Each strategy is composed of multiple approaches for addressing hazards identified in the risk assessment. These approaches focus on hazard mitigation for a specific system or collection of systems and components, as well as facility and site-related elements. Strategies may be used either singly or in combination. Past research indicates that combinations of risk mitigation strategies offer flexibility in dealing with both a single hazard and multiple hazards.

Engineering

Engineering strategies are technical options in the construction or renovation of constructed facilities, their systems, or their subsystems designed to reduce the likelihood or consequences of disasters. Engineering strategies provide protection against both natural and man-made hazards. Engineering strategies also help defend against man-made hazards, where their ability to detect or deter may reduce the likelihood or consequences of such hazards.

Protective engineering strategies are intended to reduce harm to occupants, damage to the structure, and disruption of business if a disaster occurs. Protective engineering strategies may improve the structural integrity of a building, facilitate evacuation of occupants, or circumvent compromised systems.

There is some overlap among engineering strategies that deter, detect, and protect against terrorist attacks and other criminal acts. Detection and protective engineering strategies that are observable to potential terrorists may deter them from attacking. Closed-circuit television (CCTV), for example, is designed to detect unauthorized activities, but its visibility may deter these activities.

Risk mitigation strategies may also be hazard-specific. Reinforced building shell, shatter-resistant glass, and use of barriers and bollards to achieve increased setback distances for existing buildings are examples of engineering strategies that protect against blast.

Management

Management strategies can be procedural or technical. Some management strategies relate to security, training, and communications. Others relate to decisions on where to locate the building and who should have access to its systems and subsystems. Some management strategies complement engineering strategies, while others substitute for them.

Security practices are the use of security personnel and procedures to prevent terrorist or criminal breaches from happening by detection or deterrence. They may be used to perform identification checks at building entrances, conduct background checks on individuals with access to sensitive areas and information, patrol facilities, and monitor CCTVs. Security personnel may also be used to capture attackers or facilitate recovery if a breach occurs.

Training practices are used primarily to prepare responses to disaster. Building owners and managers may institute periodic emergency response drills for building occupants. These drills may include information about evacuation routes or sheltering procedures to improve survival during emergencies. Security and facility management personnel may receive training about proper techniques for responding to breaches and containing damage. Training may also be used for prevention: building security personnel and occupants may be trained in detection of suspicious activities and notification procedures.

Building owners and managers may also use communications practices to coordinate responses with emergency personnel and to relay information and instructions to occupants during emergencies. Communications practices include setting up emergency phone numbers or instituting building-wide audio or e-mail broadcast mechanisms. Coordinated communications can play a key role in occupant safety. Building owners and managers can develop communications procedures to coordinate with first responders, security staff, and other emergency personnel responding to the incident. Finally, communications practices can be used by firms occupying the building to facilitate recovery, assess consequences, and minimize disruptions to the organization's mission or business.

Another management practice available to building owners and managers relates to the building's location and ease of access. Decisions concerning location come into play for new construction and for acquisitions of existing buildings. Setback distances, which have effects that are interdependent with some engineering strategies, are a component of the management decision about location. For new construction, managers may choose a site within a lot that satisfies a minimum setback distance. When acquiring existing property, managers may make a choice based on the physical characteristics of the available properties. Other structure-related management decisions concern access to the building itself and its sensitive areas. These access areas include attached garages, mailrooms, loading docks, side entrances, connected buildings, driveways, and rooftops. Sensitive areas include rooms housing HVAC equipment and controls; servers, network connections, and other information technology (IT) assets; and CCTV monitoring equipment.

Financial

Building owners and managers can explore financial strategies to reduce their pecuniary risks from natural and man-made hazards. There are two types of financial strategies to address risk mitigation: insurance and financial incentives.

Building owners and managers may reduce their risk exposure to disasters by purchasing insurance for worker's compensation, property damage, business interruptions, event cancellation, and liability.

Financial incentives fall into two categories: government incentives and private incentives. Government incentives are explicitly designed public policy instruments that encourage decision-makers to make certain choices over others. Private incentives reward decision-makers for making some choices over others through private transactions. In the case of risk mitigation, government and private incentives are policies, measures, or characteristics that motivate building owners and managers to implement risk mitigation measures in their buildings.

Federal, state, and local governments can institute direct incentives that reduce the price that building owners and managers pay to protect their buildings. These incentives include subsidies or tax write-offs for investments in protective measures. Other examples of government-initiated financial incentives are formal cost sharing of the protective investments and loan guarantees to ease the short-term financial burdens of structural upgrades.

Financial incentives for risk mitigation in constructed facilities may also be offered by the private sector. Building owners have commercial relationships with insurers, tenants, employees, potential buyers, and lenders. These parties may each benefit from a building's reduced vulnerability.

Insurance companies benefit from the adoption of either engineering or management strategies through smaller claims if a disaster occurs. To encourage owners to adopt risk mitigation, insurers may reduce insurance premiums for buildings that have protective measures. Building owners may also be able to obtain more favorable insurance policies, such as those that are longer term, have lower deductibles, or have fewer exclusions.

Building owners who lease commercial space may find that tenants value a building's safety features and are willing to pay a leasing premium. For owner-occupied buildings, employees may also value the added safety of a less vulnerable building. The perception of danger may affect employees' willingness to work in a particular location.

Potential buyers are another party from which a building owner can extract rewards for the building's risk mitigation measures. The installation of protective measures in a building is an improvement that increases the value of the asset. The building owner may realize the benefit of increased property value when the property is sold.

Building owners may also receive incentives from their lenders to protect their assets. Lenders would suffer direct financial losses if the destruction of a building led to the building owner's insolvency. To encourage owners to make choices that reduce the likelihood of such destruction, lenders may offer preferential financing terms on the building loan. Another way building owners are potentially rewarded in their relationships with financial institutions for their risk mitigation efforts is through the increased collateral value of their buildings.

A.2.2   Select Candidate Combinations of Risk Mitigation Strategies

Form a project team to select combinations of risk mitigation strategies. The project team will include some of the individuals from the assessment team as well as additional individuals with specific knowledge about the facility or subject matter expertise. Provide the project team with access to the compiled set of information produced by the risk assessment team.

Review the findings of the assessment team on how individual building elements are affected by each hazard type. Use information from the documents and software summarized in Chapter 3 to identify mitigation strategies for building elements and hazard types. Employ a combination of mitigation strategies rather than focusing only on engineering-based approaches.

Form each combination of risk mitigation strategies into a well-defined alternative, which addresses one or more of the hazards identified in the risk assessment. Prepare a brief narrative statement for each alternative in the set, describing what it does and how it accomplishes it (e.g., how it addresses vulnerabilities identified in the risk assessment and what implications they have for reducing consequences).

A.2.3   Develop Cost/Loss Estimates and Sequence of Cash Flows for Each Candidate Combination

Consult with senior management to establish a first cost budget constraint for the project. Compile information on the amount and timing of investment costs, operating costs, and maintenance and repair costs for each alternative combination of risk mitigation strategies. Eliminate from further consideration those alternatives whose initial investment costs exceed the first cost budget constraint for the project.

Compile information on the likelihood and consequences of each hazard type for each alternative. Develop estimated costs for each consequence.

Identify areas where information is impacted by uncertainty. Identify any significant effects that remain unquantified.

A.3 Perform Economic Evaluation

The third step in the protocol, economic evaluation, is the means through which competing alternatives are analyzed and a cost-effective risk mitigation plan is identified. The two previous steps, concerned with risk assessment and risk mitigation, formulate the alternative risk mitigation strategies and provide the associated cost and hazard data needed to compare the competing alternatives. The economic evaluation step includes the selection of the appropriate measures of economic performance, a rigorous analysis of the alternative risk mitigation strategies, the identification of the cost-effective risk mitigation plan, and the documentation necessary to support the recommendation of that plan.

Investments in long-lived projects, such as the erection of new constructed facilities or additions and alterations to existing constructed facilities, are characterized by uncertainties regarding project life, operation and maintenance costs, revenues, and other factors that affect project economics. Since future values of these variable factors are generally unknown, it is difficult to make reliable economic evaluations.

The traditional approach to uncertainty in project investment analysis is to apply economic methods of project evaluation to best-guess estimates of project input variables, as if they were certain estimates, and then to present results in a single-value, deterministic fashion. When projects are evaluated without regard to uncertainty of inputs to the analysis, decision-makers may have insufficient information to measure and evaluate the financial risk of investing in a project having a different measure of performance from what is expected.

Treatment of uncertainty and financial risk is particularly important for projects affected by natural and man-made hazards that occur infrequently, but have significant consequences. Following the three-step protocol when performing an economic evaluation assures the user that relevant economic information, including information regarding uncertain input variables, is considered for projects affected by natural and man-made hazards.

The economic evaluation step addresses uncertainty and financial risk in a structured, three-part manner. First, best-guess estimates are used to establish a baseline analysis. The baseline analysis uses fixed parameter values to calculate economic measures of performance. The results of the baseline analysis allow the alternative combinations of risk mitigation strategies to be ranked according to their economic measures of performance. The ranking of the alternatives and the calculated measures of performance provide a frame of reference for the treatment of uncertainty and financial risk. Second, a sensitivity analysis is performed in which selected inputs are varied about their baseline values. The sensitivity analysis is especially helpful in identifying shifts in the rank ordering of alternatives. The sensitivity analysis, although it addresses uncertainty in input values, produces only a crude measure of financial risk. Third, a Monte Carlo simulation is performed to obtain an explicit measure of financial risk associated with the competing alternatives. Monte Carlo simulation is especially useful in identifying shifts in the rank ordering of alternatives and documenting the factors and circumstances associated with those shifts.

A.3.1 Select Appropriate Economic Method(s) for Evaluating the Candidate Combinations of Risk Mitigation Strategies

Numerous methods are available for measuring the economic performance of investments in buildings and building systems. Use ASTM Standard Guide E 1185[10] to identify types of building design and system decisions that require economic evaluation and to match the technically appropriate economic methods with the decisions.

Four economic evaluation methods addressed in ASTM Standard Guide E 1185 apply to the development of a cost-effective risk mitigation plan for dealing with natural and man-made hazards: (1) life-cycle costs (ASTM Standard Practice E 917);[11] (2) present value net savings (ASTM Standard Practice E 1074);[12] (3) savings-to-investment ratio (ASTM Standard Practice E 964);[13] and (4) adjusted internal rate of return (ASTM Standard Practice E 1057).[14] The computer program described in reference 4.1.3.4 on page 91 produces calculated values for each of the four economic evaluation methods.

More than one method can be technically appropriate for many design and system decisions. If more than one method is technically appropriate, use all that apply, since many decision-makers request multiple measures of magnitude (life-cycle costs and present value net savings) and of return (savings-to-investment ratio and adjusted internal rate of return) to assess economic performance.

A.3.2 Compute Measures of Economic Performance for Each Candidate Combination

Follow the instructions given in the selected evaluation method(s) for computing the measure(s) of economic performance. Perform these computations with fixed parameter values. Cases where parameter values are allowed to vary are treated in Section A.3.3.

---

[10] ASTM International. "Standard Guide for Selecting Economic Methods for Evaluating Investments in Buildings and Building Systems," E 1185, *Annual Book of ASTM Standards: 2005*. Vol. 04.11. West Conshohocken, PA: ASTM International.

[11] ASTM International. "Standard Practice for Measuring Life-Cycle Costs of Buildings and Building Systems," E 917, *Annual Book of ASTM Standards: 2005*. Vol. 04.11. West Conshohocken, PA: ASTM International.

[12] ASTM International. "Standard Practice for Measuring Net Benefits for Investments in Buildings and Building Systems," E 1074, *Annual Book of ASTM Standards: 2005*. Vol. 04.11. West Conshohocken, PA: ASTM International.

[13] ASTM International. "Standard Practice for Measuring Benefit-to-Cost and Savings-to-Investment Ratios for Investments in Buildings and Building Systems," E 964, *Annual Book of ASTM Standards: 2005*. Vol. 04.11. West Conshohocken, PA: ASTM International.

[14] ASTM International. "Standard Practice for Measuring Internal Rate of Return and Adjusted Internal Rate of Return for Investments in Buildings and Building Systems," E 1057, *Annual Book of ASTM Standards: 2005*. Vol. 04.11. West Conshohocken, PA: ASTM International.

Use the computed values of the measure(s) of economic performance to rank order the alternatives (combinations of risk mitigation strategies). Refer to the selected evaluation method(s) to determine the criterion for ranking alternatives.

Designate the alternative with the best measure of economic performance as the most cost-effective risk mitigation plan. For example, if the life-cycle cost method is used, the alternative with the lowest life-cycle cost has the best measure of economic performance. Consequently, it qualifies as the most cost-effective risk mitigation plan.

Examine any significant effects that remain unquantified. Note how these effects differ across alternatives.

A.3.3 Recompute Measures of Economic Performance Taking into Consideration Uncertainty

Decision-makers typically experience uncertainty about the correct values to use in establishing basic assumptions and in estimating future costs. ASTM Standard Guide E 1369[15] recommends techniques for treating uncertainty in parameter values in an economic evaluation. It also recommends techniques for evaluating the financial risk that a project will have a less favorable economic measure of performance than what is desired or expected. ASTM Standard Practice E 1946[16] establishes a procedure for measuring cost risk for buildings and building systems, using the Monte Carlo simulation technique as described in ASTM Standard Guide E 1369. The computer program described in reference 4.1.3.4 on page 91 incorporates the treatment of financial risk and uncertainty to produce a set of calculated values for each of the four economic evaluation methods referenced in Section A.3.1 that are consistent with ASTM Standard Guide E 1369.

Perform Sensitivity Analysis

Sensitivity analysis is a test of the results of an economic evaluation to changing values of one or more parameters about which there is uncertainty. It shows decision-makers how the economic viability of a project changes as the discount rate, key unit costs, escalation rates, and other critical parameters vary.

A sensitivity analysis might use as inputs a pessimistic value, a value based on a measure of central tendency (mean or median), and an optimistic value for the parameter of interest. Then an analysis could be performed to see how each measure of performance (e.g., savings-to-investment ratio) changes as each input is considered in turn, while all other parameters are held constant. A sensitivity analysis can also be performed on

---

[15] ASTM International. "Standard Guide for Selecting Techniques for Treating Uncertainty and Risk in the Economic Evaluation of Buildings and Building Systems," E 1369, *Annual Book of ASTM Standards: 2005*. Vol. 04.11. West Conshohocken, PA: ASTM International.

[16] ASTM International. "Standard Practice Measuring Cost Risk of Buildings and Building Systems," E 1946, *Annual Book of ASTM Standards: 2005*. Vol. 04.12. West Conshohocken, PA: ASTM International.

different combinations of parameters. That is, several parameters are altered at once and then a measure of economic performance is computed.

The key advantage of sensitivity analyses is that they are easily constructed and computed and the results are easy to explain and understand. Their disadvantage is that they do not produce results that can be tied to probabilistic levels of significance (e.g., the probability that the savings-to-investment ratio is less than 1.0).

Perform Monte Carlo Simulation

Monte Carlo simulation varies a small set of key parameters either singly or in combination according to an experimental design. Associated with each key parameter is a probability distribution function from which values are randomly sampled. The major advantage of the Monte Carlo simulation technique is that it permits the effects of uncertainty to be rigorously analyzed through reference to a derived distribution of a project's measures of economic performance. Their disadvantage is that they require a computer program to implement.

In a Monte Carlo simulation, not only the expected value of the measures of economic performance can be computed but also the variability of that value. In addition, probabilistic levels of significance can be attached to the computed measures of economic performance for each alternative under consideration.

Key elements of ASTM Standard Guide E 1369 and ASTM Standard Practice E 1946 have been incorporated into the calculation of life-cycle costs (ASTM Standard Practice E 917). ASTM Standard Practice E 917 provides direction on how to apply Monte Carlo simulation when performing economic evaluations of alternatives designed to mitigate the effects of natural and man-made hazards that occur infrequently but have significant consequences. ASTM Standard Practice E 917 contains a comprehensive example on the application of Monte Carlo simulation in evaluating the merits of alternative risk mitigation strategies for a prototypical data center.

A.3.4   Analyze Results and Identify the Most Cost-Effective Combination of Risk Mitigation Strategies

Choosing among alternatives designed to reduce the impacts of natural and man-made hazards is more complicated than most building investment decisions. Consequently, guidance is provided to help identify key characteristics and the level of effort that will promote a better-informed decision. This guidance draws on information presented in Sections A.3.2 and A.3.3.

Review the calculated values of each alternative's measures of performance. Include the measures of performance computed for each of the three types of analysis: (1) fixed parameter values; (2) sensitivity analyses; and (3) Monte Carlo simulations.

Use the performance criterion from each selected evaluation method to rank order alternatives for each type of analysis (fixed parameter values, sensitivity analyses, an d Monte Carlo simulations). Document differences in alternative rankings among the three types of analysis. Focus on circumstances under which the most cost-effective risk mitigation plan identified in the fixed parameter values analysis is replaced by another alternative (other alternatives) when the effects of uncertainty are considered. Use the results of the Monte Carlo simulations to identify the characteristics associated with ranking changes for those alternatives under consideration.

Recommend an alternative as the most cost-effective risk mitigation plan. Provide a rationale for the recommendation. Include as part of the rationale, findings from each of the three types of analysis. Include a discussion of circumst ances under which the recommended alternative did not have the best measure of economic performance.

Describe any significant effects that remain unquantified. Explain how these effects impact the recommended alternative. Refer to ASTM Standard Practice E 1765[17] and its adjunct for guidance on how to present unquantified effects along with the computed values of the measures of economic performance.

A.3.5 Prepare Report with Documentation Supporting Recommended Risk Mitigation Plan

In a report of an economic evaluation, state the objective, the constraints, the alternatives considered, the key assumptions and data, and the computed value for each measure of economic performance of each alternative. Make explicit the discount rate; the study period; the main categories of cost data, including initial costs, recurring and nonrecurring costs, and resale values; and grants and incentives if integral to the decision-making process. State the method of treating inflation. Specify the assumptions or costs that have a high degree of uncertainty and are likely to have a significant impact on the results of the evaluation. Document the sensitivity of the results to these assumptions or data. Describe any significant effects that remain unquantified in the report.

Use the generic format for reporting the results of an economic evaluation described in ASTM Standard Guide E 2204.[18] It provides technical persons, analysts, and researchers a tool for communicating results in a condensed format to management and non-technical persons. The generic format calls for a description of the significance of the project, the analysis strategy, a listing of data and assumptions, and a presentation of the computed values of any measures of economic performance. ASTM Standard Guide E 2204 contains a comprehensive example evaluating the merits of alternative risk mitigation strategies for a prototypical data center summarized using the generic format.

---

[17] ASTM International. "Standard Practice for Applying the Analytical Hierarchy Process (AHP) to Multiattribute Decision Analysis of Investments Related to Buildings and Building Systems," E 1765, *Annual Book of ASTM Standards: 2005*. Vol. 04.12. West Conshohocken, PA: ASTM International.

[18] ASTM International. "Standard Guide for Summarizing the Economic Impacts of Building-Related Projects," E 2204, *Annual Book of ASTM Standards: 2005*. Vol. 04.12. West Conshohocken, PA: ASTM International.

To complete the report, include as supporting documentation information compiled from the risk assessment and a description of the process by which combinations of risk mitigation strategies were assembled.

Appendix B: Clearinghouses and Web Portals

1. **American Re.**
   URL: http://www.amre.com/

American Re is a reinsurance company and a member of the Munich Re Group, which is a prominent reinsurer. American Re provides numerous publications, articles, and web links on its website. Some of the resources are produced by American Re while others are produced by separate public and private entities. The resources from this site are categorized into topics related to the insurance industry:

Actuarial Electronic	Commerce	Healthcare/Medical Product	Recall
Agriculture	Emerging Markets	Housing Contractors	Professional Liability
Alternative Markets	Employment Liabilities	Industry Publications and News Services	Property
Arson/Fire Engineering/Technical	Risk	Insurance Association and Organizations	Public Entities
Automotive Environmental	Liability International	Media Reinsurance	
Captives Finance	And Investment	Internet Religious	
Catastrophe Financial	Institutions Legal		Risk Management
Chemicals	Financial Products	Litigation Management	Risk Management Resources
Claims	Financial Ratings	Marine (Inland Marine)	Sexual Misconduct & Abuse
Claims Issues	General Liability	Multi-Line P & C	Taxation
Economic Information	Government Statistics (US-Federal)	Market Overview	Underwriting
Educational	Government Statistics (US-States)	Nonprofit Entities	Weather/Climate
Product		Liability	Workers' Compensation

2. **CBS News. "CBS News Disaster Links."**
   URL: http://www.disasterlinks.net/

This website provided by CBS contains several hundred sites; many of the sites are discussed within this guide. WebPages and links change at a rapid rate, and as a result, a few of the links are dead. This site is categorized around the following 30 topics:

Airplane Disasters	El Nino	Oil Spills
Biological and Chemical Weapons	Emergency Management	Relief Agencies
Cyber Crime	Floods	Severe Weather
Disaster Communication	Heat	Space Weather
Disaster Education	Hurricanes	State Disaster Agencies
Disaster Imagery	Icebergs	Terrorism
Disaster Monitoring Software	International Disasters	Typhoons
Disease Landslides		Weather Maps
Drought Lightning Wind		Chill
Earthquakes Oceans Winter		Storms

3. **Center for State Homeland Security.**
   URL: http://www.cshs-us.org/

This site provides information on counterterrorism reports, homeland security readings, homeland security legislation, public health, infrastructure protection, agricultural terrorism, and a list of homeland security offices. The homeland security readings section has links to numerous news articles on such topics as the evacuation of pets during a disaster, nuclear-plant security, cargo container inspections, and hurricane Katrina. The section on homeland security legislation, pending and enacted, includes a list of relevant links. The section on counterterrorism reports is chronologically organized and dates back to October 1997. The section on public health and bio-defense readings lists news articles and other media that discusses public health and preparedness. Each article includes a short description and a link. The readings on infrastructure protection include a list a dozen articles with descriptions that cover a wide variety of topics. Examples include nuclear power plants, transit, passenger rail security, and chemical plant security. The agriculture terrorism section has about a dozen articles.

4. **Environmental Protection Agency (EPA).**
   URL: http://www.epa.gov/

See reference 2.3.2.1 (page 27)

5. **Federal Emergency Management Agency (FEMA).**
   URL: http://www.fema.gov/

See Reference 2.1.6 (page 9)

6. **Federal Emergency Management Agency. Catalog of FEMA Earthquake Publications. Washington DC: FEMA, April 2006.**
   URL: http://www.nehrp.gov/info/PDF/NehrpCatalog2006-FEMA2.pdf

This publication lists all the FEMA Earthquake Publications. There are eighteen pages of publications; most of them are available online.

7. **Federal Emergency Management Agency. "Risk Management Series Publications."**
   URL: http://www.fema.gov/plan/prevent/rms/index.shtm

FEMA's Risk Management Series publications (see reference 3.1.21 through 3.1.34 on pages 38 through 42), which are available on FEMA's website, are directed at providing design guidance for mitigating terrorist risks. The objective of the Risk Management Series is to reduce physical damage to structural and nonstructural components of buildings and related infrastructure, and to reduce casualties resulting from bomb, chemical, biological, and radiological attacks. The emphasis of this series is to improve security in high occupancy buildings to better protect the nation from potential threats by identifying actions and designs to strengthen buildings.

8. **Government Printing Office. "GPO Access."**
   URL: http://www.gpoaccess.gov/

The U.S. Government Printing Office (GPO) provides information from all three branches of the government to the public. Using the GPO website one can identify departments dealing with terrorism, telecommunications, statistics, radiation, construction, and a number of other topics (http://www.gpoaccess.gov/topics/index.html). This site serves as a gateway to all federal agencies and departments.

9. **High Plains Regional Climate Center and University of Nebraska. "Climate and Weather: Data, Information, and Products Clearinghouse."**
   URL: http://www.hprcc.unl.edu/clearinghouse/index.html

See Reference 2.3.1.5 (page 21)

10. **Institute for Business and Home Safety (IBHS).**
    URL: http://www.ibhs.org

See reference 3.1.41 (page 44)

11. **International Strategy for Disaster Reduction (ISDR).**
    URL: http://www.unisdr.org/

See reference 3.1.44 (page 45)

12. **Marlatt, Greta E. Chemical, Biological, and Nuclear Terrorism/Warfare: A Bibliography. Dudley Knox Library: Naval Postgraduate School, September 2003.**
    URL: http://library.nps.navy.mil/home/bibs/chemtoc.htm

See reference 2.3.2.8 (page 28)

13. **Munich Re Group.**
    URL: http://www.munichre.com/

See Reference 3.3.2.10 (page 66)

14. **NASA. "Natural Disaster Reference Database."**
    URL: http://ndrd.gsfc.nasa.gov/

This site lists a large number of articles by natural hazard topic: wildfire, tsunami, avalanche, eruptions, earthquake, landslide, flooding, hurricane, storm surge, tornadoes, cyclones, drought, typhoon, disease, and natural disasters in general. These articles are technical in nature and discuss the details of weather modeling. The website provides the title, author, and source of the article along with an abstract, but there is no link to the article.

15. **National Earthquake Hazards Reduction Program (NEHRP).**
    URL: http://www.nehrp.gov/

See reference 3.1.55 (page 48)

16. **National Fire Protection Association (NFPA).**
    URL: http://www.nfpa.org

See reference 2.3.2.9 (page 28)

17. **National Institute of Standards and Technology (NIST): Building and Fire Research Laboratory.**
    URL: http://www.bfrl.nist.gov

See reference 2.2.10 (page 18)

18. **National Memorial Institute for the Prevention of Terrorism (MIPT).**
    URL: http://www.mipt.org/

See reference 3.3.2.11 (page 67)

19. **National Oceanic and Atmospheric Administration (NOAA).**
    URL: http://www.noaa.gov/

See reference 2.3.1.10 (page 22)

20. **RAND Corporation. May 8, 2006.**
    URL: http://www.rand.org/

See reference 2.1.21 (page 13)

21. **Risk Management Solutions Incorporated (RMS).**
    URL: http://www.rms.com/

See reference 2.1.22 (page 14)

22. **Swiss Re.**
    URL: http://www.swissre.com/

See Reference 3.3.2.22 (page 69)

23. **Tec-Com Incorporated. RiskWorld.**
    URL: http://www.riskworld.com/

See Reference 2.2.18 (page 20)

24. **United States Department of Health and Human Services. "Disasters and Emergencies."**
    URL: http://www.hhs.gov/disasters/index.shtml

The "Disasters and Emergencies" website is developed by the U.S. Department of Health and Human Services and provides a number of links to other departments and agencies related to the following disaster categories: biological, chemical, and radiological weapons; bioterrorism, disasters and emergencies; environmental disasters; homeland security; natural disasters/extreme weather; mental health and traumatic events; safety of the water supply; and other resources.

25. **United States Department of Homeland Security (DHS).**
    URL: http://www.dhs.gov

See Reference 3.1.75 (page 53)

26. **United States Department of Homeland Security. "Homeland Security Digital Library."**
    URL: https://www.hsdl.org

The Department of Homeland Security (DHS) established the Homeland Security Digital Library (HSDL), which is restricted to researchers and government employees. To gain access to the library visit the webpage and apply for a password. DHS will probably respond within a few days. This extensive database contains reports, articles, and other publications that can be browsed by topic or searched using a basic or advanced search option. Most of the entries in the HSDL have a link to view the full publication.

27. United States Geological Survey (USGS).
   URL: http://www.usgs.gov/

See Reference 2.3.1.20 (page 25)

28. United States Geological Survey: National Geospatial Program Office.
   URL: http://www.usgs.gov/ngpo/

See Reference 2.3.1.23 (page 25)

29. University of Colorado-Boulder Natural Hazards Center.
   URL: http://www.colorado.edu/hazards/resources/sites.html

This website provides an annotated index of useful disaster web sites. Each entry includes the web address and a brief description of the content from the site. They are grouped into eighteen categories:

All hazards	Email Lists	Severe Weather
Career/Employment	Floods	Snow Avalanche
Climate Change	Hurricanes	Space Hazards
Data and Costs	Images	Tsunamis
Disaster Health	landslides	Volcanoes
Earthquakes	Satellites Wildfire	

Appendix C: Policies, Research, and Theory

1. **Arrow, Kenneth J. "The Theory of Risk-Bearing: Small and Great Risks." Journal of Risk and Uncertainty. 12 (1996): 103-111.**

This article discusses risk-sharing and insurance as mutually advantageous transactions. It explores the ideal model of insurance versus real world insurance and offers some explanations.

2. **Boadway, Robin, Maurice Marchand, Manuel Leite-Monteiro, and Pierre Pestieau. "Social Insurance and Redistribution with Moral Hazard and Adverse Selection." CORE Discussion Papers. 2004/83. December 7, 2004.
URL: http://www.core.ucl.ac.be/services/psfiles/dp04/dp2004_83.pdf**

This paper studies "how equity and efficiency considerations should be traded off in choosing the optimal coverage of social insurance." The study considers the influence of ex post moral hazard, adverse selection, individual's effect on event loss, and private supplements to social insurance.

3. **Bush, George W. "Homeland Security Presidential Directives."
URL: http://www.whitehouse.gov**

The *Homeland Security Presidential Directives* (HSPD) record and communicate security decisions by the president. To date, there are thirteen directives. In addition to the White House website, these documents can be accessed at http://www.fbiic.gov/executiveorders.htm and through the Government Printing Office website http://www.gpoaccess.gov/. HSPD-1 outlines the functions and organization of the Homeland Security Council. HSPD-2 establishes the Foreign Terrorist Tracking Task Force and updates immigration policies concerning the INS, international students, and data sharing. The purpose of HSPD-3 is to establish the Homeland Security Advisory System, which uses a color coded scheme to represent the current level of threat to U.S. security. HSPD-4 is classified, but an unclassified version is available to the public. It discusses the governments approach to countering weapons of mass destruction. In HSPD-5 the president enhances the response to domestic incidences and establishes a national incident management system. HSPD-6 outlines the integration and use of screening information. The purpose of HSPD-7 is to establish a national policy to identify and prioritize critical infrastructure and resources. HSPD-8 establishes a national domestic "all-hazards preparedness goal" for improved delivery of federal assistance to state and local governments. HSPD-9 develops a national policy to protect the agriculture and food system against natural and man-made disasters. Biological weapons are discussed and addressed in HSPD-10. The subject of HSPD-11 is the "comprehensive terrorist-related screening procedures." HSPD-12 establishes a common standard for the identification of federal employees, and HSPD-13 provides guidelines in order to protect U.S. maritime interests.

4. **Daniels, Ronald J., Donald F. Kettl, and Howard Kunreuther. On Risk and Disaster. Philadelphia, PA: University of Pennsylvania Press, 2006.**

This text provides a discussion of the events caused by hurricane Katrina and raises questions concerning risk and the response to disasters in the context of public and private roles. The text provides expert insight concerning better preparedness, mitigation, and response to disasters. Part one of the text discusses the gulf coast and its future. Part two discusses risk management, and part three discusses private sector risk management strategies. The last section, part four, explores the government's role in risk management and disaster response.

5. **Environmental Systems Research Institute. GIS for Homeland Security. Redlands, CA: ESRI Press, 2005.**

This booklet discusses the use of geographic information systems (GIS) as a tool for homeland security. It discusses decision making, threat identification, response, preparedness, agriculture, infrastructure, and other GIS related topics.

6. **Glickman, Theodore S. and Michael Gough. Readings in Risk. Washington DC: Resources for the Future, 1995.**

This text is a collection of articles on environmental and technological risks. It discusses health risk assessment, technological risk assessment, and risk communication. A number of the articles are listed below.

- Application of Risk Assessment to Food Safety Decision Making
- Risk, Science, and Democracy
- Cost-Benefit Analysis: An Ethical Critique
- Rating the Risks
- Ranking Possible Carcinogenic Hazards
- Social Benefit Versus Technological Risk
- Assessing Risks from Health Hazards: An Imperfect Science
- The Emergence of Risk Communication Studies: Social and Political Context

7. **Godschalk, David R., Timothy Beatley, Philip Berke, David J. Brower, Edward J. Kaiser, Charles C. Bohl, and R. Matthew Goebel. Natural Hazard Mitigation. Washington, DC: Island Press, 1999.**

This text offers recommendations for reforming federal and state strategies that concern natural hazard mitigation. It discusses several natural disasters and the actions that were taken by the local, state, and federal governments to mitigate them. Within the text are tables and discussions of current disaster funding and a review of current policies.

8. Government Accountability Office. Catastrophe Risk: U.S. and European Approaches to Insure Natural Catastrophe and Terrorism Risks. GAO-05-199. Washington DC: GAO, February 2005.
   URL: http://www.gao.gov/new.items/d05199.pdf

This publication provides an overview of the insurance industry and its capacity to cover low probability high cost events. It analyzes catastrophe bonds and discusses the approaches that six European countries have used for natural hazard and terrorist risks.

9. Government Accountability Office. Risk Management: Further Refinements Needed to Assess Risks and Prioritize Protective Measures at Posts and other Infrastructure. GAO-06-91. Washington DC: U.S. Government Accountability Office, December 2005.
   URL: http://www.gao.gov/new.items/d0691.pdf

This publication examines the Coast Guard, the Office for Domestic Preparedness, and the Information Analysis and Infrastructure Protection Directorate and the measures they have taken to protect the nation against threats. It also makes a number of recommendations as to how these entities can improve protection.

10. Hamm, Mark S. Crimes Committed by Terrorist Groups: Theory, Research, and Prevention. Document 211203. Award 2003-DT-CX-0002.
    URL: http://www.ncjrs.gov/pdffiles1/nij/grants/211203.pdf

This research report examines major terrorist trials in order to identify the techniques and attributes of various terrorist organization types. It shows that terrorist-oriented organizations have distinguishing features that can be identified by conventional criminal investigations. The report could be useful to any organization that might come in contact with terrorist group activities (e.g., financial institutions, law enforcement, and security).

11. Heal, Geoffrey, and Howard Kunreuther. "You Can Only Die Once: Public-Private Partnerships for Managing the Risks of Extreme Events." Risk Management Strategies in an Uncertain World (Conference). New York: April 12-13, 2002.
    URL: http://www2.gsb.columbia.edu/faculty/gheal/EconomicTheoryPapers/index.html

This paper explores private investment in security in the context of others behavior and discusses security investment from a social and private perspective. The question that motivates Heal and Kunreuther is whether "organizations, such as airline companies and computer network managers, invest in security to a degree that is adequate from either a private or social perspective."

12. Jetter, James, and David Proffitt. "Effectiveness of Expedient Sheltering in Place in Commercial Buildings." Journal of Homeland Security and Emergency Management. Vol. 3, No. 2, Article 4. 2006.
    URL: http://www.bepress.com/jhsem/vol3/iss2/4/

This article examines the effectiveness of shelter in place in commercial buildings for protection against airborne hazards. It makes comparisons of leaky, typical, and tight buildings and shelters for various scenarios.

13. Kunreuther, Howard. "Mitigating Disaster Losses through Insurance." Journal of Risk and Uncertainty. Vol. 12, Numbers 2-3 (May 1996): 171-187.

This article discusses the reasons why homeowners have not voluntarily adopted cost-effective measures for disaster mitigation. It then suggests that insurance and well-enforced building codes be used to mitigate future damage. According to this article, financial institutions and insurance companies play a key role in mitigation.

14. Kunreuther, Howard, Erwann Michel-Kerjan, and Beverly Porter. "Assessing, Managing and Financing Extreme Events: Dealing with Terrorism." National Bureau of Economic Research. Working Paper 10179 (December 2003).
    URL: http://www.nber.org/papers/w10179

This paper discusses risk assessment, risk management, and risk financing as they relate to catastrophic terrorist risk. It explores the implications of the Terrorism Risk Insurance Act (TRIA) and models for assessing risk for establishing premiums. International experiences with terrorism insurance and successful terrorism insurance programs are discussed.

15. Kunreuther, Howard, and Mark Pauly. "Neglecting Disaster: Why Don't People Insure Against Large Losses." Journal of Risk and Uncertainty. Vol. 28, Number 1 (January 2004): 5-21.

This paper discusses why people do not purchase insurance with favorable premiums for low-probability high-loss events. It proposes that "individuals maximize expected utility but face an explicit or implicit cost to discovering the true probability of rare events," which inhibits the purchase of insurance.

16. Kunreuther, Howard, Nathan Novemsky, and Daniel Kahneman. "Making Low Probabilities Useful." Journal of Risk and Uncertainty. Vol. 23, Number 2 (September 2001): 103-120.

This paper discusses the perception of low probability-high consequence events. It concludes that a significant amount of information must be provided for a layperson to determine the difference between low probability events. One needs to present comparison scenarios that are located on a probability scale to evoke people's feelings of risk.

17. Marshall, Harold. "Economic Approaches to Homeland Security for Constructed Facilities." Tenth Joint W055-W065 International Symposium on Construction Innovation and Global Competitiveness. September 11, 2002.
    URL: http://www.bfrl.nist.gov/oae/publications/proceedings/CIBKeynoteAddress.pdf

This article presents economic models to select optimal protective strategies against terrorist-induced damages. It uses life-cycle cost and net savings models in order to choose the cost-effective level of investment. This article also provides a method to select the optimal combination of protective strategies.

18. McCarthy, John A. Critical Infrastructure Protection Program: Workshop II Working Papers. Fairfax, VA: George Mason University Press, 2004.
    URL: http://cipp.gmu.edu/research/Workshop2WorkingPapers.php

This text includes numerous working papers on critical infrastructure protection. The website listed above provides a link that has the titles to all the papers. Topics of the papers include public policy, power grids and internet networks, cyber security, wireless telecommunications, terrorism scenarios for water distribution, indoor air quality alert systems, risk assessment, and terrorism insurance. There are over 35 articles in this text.

19. National Capital Planning Commission. Designing for Security in the Nation's Capital. Washington DC: NCPC, October 2001.
    URL: http://www.ncpc.gov/planning_init/security/DesigningSec.pdf

This text describes the interim security measures that were necessary following September 11, 2001. It discusses design solutions for Pennsylvania Avenue and its impact on historic resources, traffic, and the downtown economy.

20. National Capital Planning Commission. National Capital Urban Design and Security Plan. Washington DC: NCPC, October 2002.
    URL: http://www.ncpc.gov/publications_press/NCUDSP.html

The National Capital Planning Commission has developed this design strategy to address necessary security measures in the nation's capital. Currently, security is being addressed through improvised defense measures that have had a negative impact on the life and society of Washington DC. This text focuses primarily on perimeter building security and does not address other kinds of security measures. The purpose of this plan is to "restore the beauty and dignity of the Nation's Capital by integrating building perimeter security into an attractive streetscape and by coordinating the design and installation of streetscape projects."

21. **National Consortium for the Study of Terrorism and Responses to Terrorism (START).**
    URL: http://www.start.umd.edu/

The National Consortium for the Study of Terrorism and Responses to Terrorism (START) is a Department of Homeland Security Center of Excellence, which aims at exploring the methods and resources to understand terrorism and its risks. A number of articles, books, book chapters, conference papers, and reports are listed on their website.

22. **National Institute of Standards and Technology. Performance of Physical Structures in Hurricane Katrina and Hurricane Rita: A Reconnaissance Report. NIST Technical Note 1476. Gaithersburg, MD: NIST, June 2006.**
    URL: http://www.bfrl.nist.gov/investigations/pubs/NIST_TN_1476.pdf

This report presents the findings that came from reconnaissance following Hurricanes Katrina and Rita. It discusses environmental effects (e.g., wind speed, storm surge, flooding), building damage, infrastructure damage (including levee and floodwall damage), and damage to residential structures caused by the two hurricanes. The report identifies three areas of study that need to be addressed: "(1) evaluate the performance of the New Orleans flood protection system and provide credible scientific and engineering information for guiding the immediate repair and future upgrade of the system; (2) develop risk-based storm surge maps for use in flood-resistant design of structures, and (3) evaluate and, if necessary, modify the Saffir-Simpson hurricane scale's treatment of storm surge effects due to hurricanes."

23. **National Institute of Standards and Technology. Final Report on the Collapse of the World Trade Center Towers. NIST NCSTAR 1. Gaithersburg, MD: NIST, September 2005.**
    URL: http://wtc.nist.gov/NISTNCSTAR1CollapseofTowers.pdf

This report details and reconstructs the collapse of the World Trade Center towers. It details the construction of the towers, the materials, and the events that caused their collapse. The report provides NIST's findings from the investigation of 30 recommendations in eight areas to improve building safety: structural integrity, fire endurance of structures, fire resistant design of structures, improved active fire protection, improved building evacuation, improved emergency response, improved procedures and practices, and education and training.

24. **National Science Foundation.**
    URL: http://www.nsf.gov/

The National Science Foundation is a federal agency created "to promote the progress of science; to advance the national health, prosperity, and welfare; [and] to secure the national defense…" Their website has a number of publications and statistics that are available to view over the internet. Publications include a large variety of topics.

Statistics include the following categories: Education, federal government, industry, international, research and development, social dimensions, state, and workforce.

25. **Organization for Economic Co-operation and Development. "Economic Consequences of Terrorism."**
    URL: http://www.oecd.org/dataoecd/11/60/1935314.pdf

This paper discusses and analyzes the effects of terrorism on the economy and the effect of policies that were developed following the September 11 attack. It purports that "good crisis management helped restore confidence," but some policies could have "detrimental economic consequences" in the future. This paper has a number of facts and figures, but tends to be non-technical in nature.

26. **Pate-Cornell, M. Elisabeth. "Global Risk Management." Journal of Risk and Uncertainty. 12 (1996): 239-255.**

This article discusses risk management strategies based on probability risk analysis and other management factors for industries that deal with hazardous systems. It compares the use of insurance versus actual risk reduction and explores the implications of several case study illustrations: tiles of the U.S. space shuttle, offshore platforms, marine pipelines, and anesthesia in modern hospitals.

27. **Priest, George L. "The Government, the Market, and the Problem of Catastrophic Loss." Journal of Risk and Uncertainty. 12 (1996): 219-237.**

This article discusses comparative advantage in regards to the government and private industry providing insurance for catastrophic losses. The article argues that the government is less effective than the private insurance market in providing coverage of losses.

28. **Smith, V. Kerry, and Daniel G. Hallstrom. Designing Benefit-Cost Analyses for Homeland Security Policies. EMW-2004-GR-0112. Washington DC: United States Department of Homeland Security, August 11, 2004.**
    URL: https://www.hsdl.org/

This paper discusses the methodological and empirical issues concerning benefit-cost analyses for homeland security. It first describes the measurement of the economic value of policies that reduce risk, and then it uses "parallel" hazards as a source of information for behavioral responses to large-scale disasters.

29. **Stern, Paul C. and Harvey V. Fineberg. Understanding Risk: Informing Decisions in a Democratic Society. Washington, DC: National Academy Press, 1996.**

This text provides methods to improve communication between researchers and decision-makers. It illustrates that translating scientific information is not sufficient for characterizing risks to decision-makers. "Risk Characterization should be a decision-driven activity, directed toward informing choices and solving problems… [It] is the outcome of an analytic-deliberative process."

30. **United States Department of Homeland Security. Nation Wide Plan Review: Phase 2. Washington DC: United States Department of Homeland Security, June 16, 2006.**
    URL: http://www.dhs.gov/interweb/assetlibrary/Prep_NationwidePlanReview.pdf

This publication reviews the emergency plans of all the states and 75 urban areas across the nation. It found that although there is renewed emphasis on planning and many new initiatives current planning is "unsystematic and not linked within a national planning system." It recommends a "modernization of our nation's planning processes" is essential for an adequate response to catastrophe.

31. **United States Department of Homeland Security. National Infrastructure Protection Plan: Base Plan. January 2006.**
    URL: www.dhs.gov/xlibrary/assets/NIPP_Plan.pdf

The National Infrastructure Protection Plan (NIPP) is intended to protect the nation's critical infrastructure and key resources through deterrence, mitigation, response, and prevention. The NIPP provides a unifying structure for existing protection efforts to be unified. The program focuses on three objectives: building security partnerships, implementing a long-term risk reduction program, and optimizing protection resources.

32. **United States Department of Homeland Security. National Response Plan. December 2004.**
    URL: http://www.dhs.gov/interweb/assetlibrary/NRP_FullText.pdf

The National Response Plan was conceived in order to establish a comprehensive all-hazards approach to domestic incidents, which includes prevention, preparedness, response, and recovery. This document discusses planning; the role of local, state, and federal government; the role of the private sector; the operations process; incident management; and plan maintenance/management.

33. **United States Department of Homeland Security. Securing Our Homeland: U.S. Department of Homeland Security Strategic Plan. 2004.**
    URL: http://www.dhs.gov/xlibrary/assets/DHS_StratPlan_FINAL_spread.pdf

This illustrated document provides the vision and goals for the Department of Homeland Security. The stated mission of the department is to "lead the unified national effort to secure America… prevent and deter terrorist attacks… protect against and respond to threats and hazards to the nation… ensure safe and secure borders, welcome lawful immigrants and visitors, and promote the free-flow of commerce."

34. United States Department of the Treasury: Office of Economic Policy. "Assessment: The Terrorism Risk Insurance Act of 2002." Washington, DC: United States Department of the Treasury, June 30, 2005.
    URL: http://www.ustreas.gov/press/releases/reports/063005%20tria%20study.pdf

The Terrorism Risk Insurance Act of 2002 required the Department of the Treasury to provide compensation for losses covered by insurers that were incurred by terrorists. The act is intended to protect consumers, address market disruptions, and stabilize the insurance market. This document discusses the effectiveness of the Terrorism Risk Insurance Act of 2002 to accomplish those goals and evaluates the ability of the insurance industry to insure damages caused by terrorists.

35. United States Geological Survey. The Plan to Coordinate NEHRP Post-Earthquake Investigations. Circular 1242. Reston, VA: U.S. Geological Survey, 2003.
    URL: http://geopubs.wr.usgs.gov/circular/c1242/c1242.pdf

This report is a plan to coordinate post-earthquake investigations that are conducted through the National Earthquake Hazards Reduction Program (NEHRP). A coordinated effort can improve emergency response, create a better understanding of earthquake hazards, improve land use, and facilitate a more cost effective and safer built environment.

36. White House. National Security Strategy of the United States. March 2006.
    URL: http://www.whitehouse.gov/nsc/nss/2006/

This White House publication outlines the national security strategy and discusses public policies within the U.S. government. The content is general in nature and covers international policy, regional conflicts, weapons of mass destruction, opening societies, and globalization.

37. Woo, Gordon. "Quantifying Insurance Terrorism Risk." Risk Management Solutions. February 1, 2002.
    URL: http://www.rms.com/NewsPress/Quantifying_Insurance_Terrorism_Risk.pdf

This article discusses the quantification of terrorism risk in order for the insurance market to provide coverage. It explores the different structures of terrorist organizations along with the frequency of planned attacks.

38. Woo, Gordon. "Quantitative Terrorism Risk Assessment." Risk Management Solutions. 2004.
    URL: http://www.rms.com/NewsPress/Quantitative_Terrorism_Risk_Assessment.pdf

This paper discusses the quantification of terrorism risk. Topics include the likelihood of attack, loss severity distribution, the Markov Model, risk analysis, and risk transfer. This paper is relatively technical in nature.

39. Zeckhauser, Richard. "The Economics of Catastrophes." Journal of Risk and Uncertainty. 12 (1996): 113-140.

This article explores the exacerbation of disasters by human behavior along with the dissemination of information and the response to disasters. It discusses micromotives, public policies, catastrophe prevention, insurance, and liability.

GLOSSARY OF RELATED TERMS

This glossary is a collection of economic terms related to evaluating the cost-effectiveness of disaster mitigation investments in constructed facilities. Many of these terms are used in the Cost Effectiveness Tool in reference 4.1.3.4.

100 Year Flood (Base Flood): A flood that has a 1 percent chance of occurring in any given year. A 500 Year Flood has 0.2 percent chance of occurring in any given year.

Adjusted Internal Rate of Return (AIRR): The average annual yield from a project over the study period, taking into account reinvestment of interim receipts. The reinvestment rate in the AIRR calculation is equal to the discount rate.

AIRR: See Adjusted Internal Rate of Return.

Annual Value: A uniform annual amount equivalent to the project costs or benefits taking into account the time value of money throughout the study period. Life-cycle costs may be expressed in either annual value terms or present value terms.

Annually Recurring: Means of classifying/allocating costs that occur every year within the frequency of occurrence choices of the CET software for O&M or other cost items. The three occurrence frequency choices are *annually recurring*, periodic (other than annual), and aperiodic.

Aperiodic: Means of classifying/allocating costs that follow an irregular schedule (not strictly periodic) within the frequency of occurrence choices of the CET software for O&M or other cost items. The three occurrence frequency choices are annually recurring, periodic (other than annual), and *aperiodic*.

Asset: Any man-made or natural feature that has value. This can include infrastructure, buildings, people, parks, and many other items.

Baseline Analysis: The starting point for conducting an economic evaluation. In the baseline analysis, all data elements entering into the calculations are fixed. The term baseline analysis is used to denote a complete analysis in all respects but one; it does not address the effects of uncertainty.

Base Year (Time): The date to which all future and past benefits/costs are converted when a present or annual value method is used.

Building/Facility Elements: One of the three cost types that define the building/facility component of the detailed cost-accounting framework: *building/facility elements*; building/facility site work; non-elemental. The *building/facility elements* cost type is associated with the elemental classification UNIFORMAT II.

Capital Investment: One of the three cost types that define the budget category classification: *capital investment*, O&M (operations and maintenance), other. The cost of acquiring, substantially improving, expanding, changing the functional use of, or replacing a building or building system. *Capital investment* costs accrue to the investment cost category, while O&M and other costs accrue to the non-investment cost category.

Cash Flow: The stream of monetary (dollar) values – benefits and costs – resulting from a project investment.

Constant Dollar Analysis: Dollars of uniform purchasing power exclusive of general inflation or deflation; based on the value of a dollar in a specified base year.

Constructed Facilities: Permanent structures, including infrastructure, buildings, and industrial facilities.

Cost-Accounting Framework: Methodology for tracking how costs affect stakeholders in different ways. The cost-accounting framework promotes a detailed, consistent breakdown of life-cycle costs.

Cost Effective: The condition whereby the present value benefits (savings) of an investment exceed its present value costs.

Current Dollar Analysis: Analysis of the costs incurred in dollars of purchasing power in which actual prices are stated (not corrected for inflation or deflation).

Disaster Mitigation: Measures, procedures, and strategies designed to reduce either the likelihood or consequences of a disaster.

Discount Rate: The rate of interest reflecting the investor's time value of money, used to determine discount factors for converting benefits and costs occurring at different times to a base year. The discount rate may be expressed as nominal or real.

Escalation Rate: The rate of change in price for a particular good or service (as contrasted with the inflation rate, which is for all goods and services).

Externality: The discrepancy between private and social costs or private and social benefits.

Financial Mechanisms: One of the three mitigation strategy classifications (engineering alternatives; management practices; *financial mechanisms*). A set of devices relating to finances that facility owners and managers can utilize to reduce their exposure to natural and man-made hazards. These devices include purchase of insurance policies and responding to external financial incentives to engage in engineering-based or management-based risk mitigation.

First (Initial) Costs: Attribute of a capital investment. Costs incurred in placing a building or building subsystem into service, including, but not limited to, costs of planning, design, engineering, site acquisition and preparation, construction, purchase, installation, property taxes and interest during the construction period, and construction-related fees.

GIS: See Geographic Information System

Geographic Information System (GIS): A computer system that integrates, stores, edits, and analyzes geographic information.

Inflation: A rise in the general price level over time, usually expressed as a percentage rate.

Investment Cost: First cost and later expenditures which have substantial and enduring value (generally more than one year) for upgrading, expanding, or changing the functional use of a building or building system.

LCC: See Life-Cycle Cost.

Life-Cycle Cost (LCC): A technique of economic evaluation that sums over a given study period the costs of initial investment (less resale value), replacements, operation (including energy use) and maintenance of an investment decision. Life-cycle costs may be expressed in either present value terms or annual value terms.

Mitigation Strategy: One of the four core components of the cost-accounting framework (bearer of costs; budget category; building/facility component; *mitigation strategy*). Means of classifying/allocating costs within the CET software in regards to risk management. The three *mitigation strategy* classifications include engineering alternatives; management practices; financial mechanisms.

Monte Carlo Simulation: A technique used to evaluate models that are too complicated for an analytical solution. It involves the use of numerous trials to find the equilibrium of a sytem.

Nominal Discount Rate: The rate of interest reflecting the time value of money stemming both from inflation and the real earning power of money over time. This is the discount rate used in discount formulas or in selecting discount factors when future benefits and costs are expressed in current dollars.

Operating Cost: The expenses incurred during the normal operation of a building or a building system or component, including labor, materials, utilities, and other related costs.

Present Value: The value of a benefit or cost found by discounting future cash flows to the base year. Life-cycle costs may be expressed in either present value terms or annual value terms.

Present Value Net Savings (PVNS): A method for finding the economically efficient choice among investment alternatives. It measures the net savings from investing in a given alternative instead of investing in the foregone opportunity (e.g., some other alternative or the base case). The PVNS for a given alternative, A_j, vis-à-vis the base case, A_0, may be expressed as: $PVNS_{j:0} = LCC_0 - LCC_j$

Probability of Occurrence: Provides the chance that a there will be a specific outcome associated with a given event. The sum of all outcome probabilities for a single event must be equal to 1. Listed as one of the three key parameters for a given outcome: *probability of occurrence*, first year, and last year.

PVNS: See Present Value Net Savings.

Real Discount Rate: The rate of interest reflecting that portion of the time value of money related to the real earning power of money over time. This is the discount rate used in discount formulas or in selecting discount factors when future benefits and costs are expressed in constant dollars.

Replacement Costs: Building component replacement and related costs, included in the capital budget, that are expected to be incurred during the study period.

Resale Value: The monetary sum expected from the disposal of an asset at the end of its economic life, its useful life, or at the end of the study period.

Retrofit: The modification of an existing building or facility to include new systems or components.

Risk Analysis: The body of theory and practice that has evolved to help decision-makers assess their risk exposures and risk attitudes so that the investment that is "best for them" is selected.

Risk Mitigation: The actions or decisions designed to reduce the financial and nonpecuniary risk from uncertain events.

Salvage Value: The value of an asset, assigned for tax computation purposes, that is expected to remain at the end of the depreciation period (represented as a negative cost value). One of three time classification attributes of a capital investment: initial, future, salvage.

Savings-to-Investment Ratio (SIR): Either the ratio of present value savings to present value investment costs, or the ratio of annual value savings to annual value investment costs.

Sensitivity Analysis: A means for addressing the effects of uncertainty. A test of the outcome of an analysis by altering one or more parameters (key data elements or input variables) from (an) initially assumed value(s). A *sensitivity analysis* complements the *baseline analysis* by evaluating the changes in output measures when selected data inputs are allowed to vary about their baseline values.

SIR: See Savings-to-Investment Ratio.

Study Period: The length of time over which an investment is analyzed.

UNIFORMAT II: An elemental format based on major components common to most buildings. It serves as a consistent reference for analysis, evaluation, and monitoring of buildings during the planning, feasibility, and design stages. It also enhances reporting at all stages in construction. The two cost types, building/facility elements and building/facility site work, under the building/facility component cost classification are associated with the elemental classification *UNIFORMAT II*. Subcategories under *UNIFORMAT II* include: substructure, shell, interiors, services, equipment & furnishings, special construction/demolition.

Author Index by Reference Number

4Clicks-Solutions. 3.4.1
Advanced materials and Processes Technology Information Analysis Center. 3.1.1
AIR. 2.2.2
Air Dispersion Modeling Incorporated. 2.2.1
Aldy, Joseph. 3.3.2.29
American Institute of Architects. 3.1.2
American Management Association. 3.1.3
American National Standards Institute. 4.1.2.1
American Re. Appendix B:1
American Red Cross. 3.1.4
American Society for Industrial Security. 3.1.5, 2.1.1
American Society for Testing and Materials. 4.1.2.24, 4.1.2.23, 4.1.2.22, 4.1.2.21, 4.1.2.20, 4.1.2.19, 4.1.2.18, 4.1.2.17, 4.1.2.16, 4.1.2.15, 4.1.2.14, 4.1.2.13, 4.1.2.11, 4.1.2.10, 4.1.2.9, 4.1.2.8, 4.1.2.7, 4.1.2.6, 4.1.2.5, 4.1.2.4, 4.1.2.3, 4.1.2.2, 2.1.2
American Society of Civil Engineers. 4.1.1.25, 3.1.71, 3.1.36, 3.1.17, 3.1.10, 3.1.9, 3.1.8, 3.1.7, 3.1.6, 2.1.3
American Society of Mechanical Engineers. 2.1.4
American Society of Professional Estimators. 4.1.2.28
Amos, Scott. 3.3.1.2
Arrow, Kenneth J. Appendix C:1
Association for the Advancement of Cost Engineering. 3.3.1.3, 3.3.1.2
Asvaco. 2.2.3
Autodesk. 3.2.2
Avila, Ernesto A. 3.1.17
BALFOUR Technologies LLC. 3.2.3
Bartol, Nadya. 3.1.38
Beatley, Timothy. Appendix C:7
Bentley Systems. 3.2.4
Berke, Philip. Appendix C:7
Bid2Win. 3.4.2
Bledsoe, John D. 3.3.1.4
Blue Ribbon Panel on Bridge and Tunnel Security. 3.1.11
BMS Solutions. 3.2.5
Boadway, Robin. Appendix C:2
Bohl, Charles C. Appendix C:7
BOMA International. 3.3.1.5
British Standards Institute. 4.1.2.29
Brower, David J. Appendix C:7
Building Seismic Safety Council. 3.1.12
Bureau of Justice Assistance. 2.1.18
Bureau of Labor Statistics. 3.3.2.1
Bush, George W. Appendix C:3
Carson, Richard T. 3.3.2.5
Cauffman, Stephen. 3.1.13
CBS News. Appendix B:2
Center for Risk and Economic Analysis of Terrorism Events. 4.1.1.14
Center for State Homeland Security. Appendix B:3
Centers for Disease Control National Center for Health Statistics. 3.3.2.2
Chapman, Robert E. 4.1.1.16, 4.1.1.15, 3.1.47
Chemical Safety Board. 2.3.2.1

Construction Engineering Research Laboratory. 3.3.2.8
Construction Industry Institute. 3.1.14
Construction Management Economics. 3.3.1.6
Construction Management Software. 3.4.3
Construction Specifications Institute. 4.1.2.30
Daniels, Ronald J. Appendix C:4
Decisioneering. 2.2.4
Defense Threat Reduction Agency. 3.2.6, 2.2.5
Deisler, Paul F., Jr. 2.1.5
Design Cost Data. 3.3.1.7
Earthquake Engineering Research Institute. 3.1.16, 3.1.15, 2.3.1.1
EarthQuake Information NETwork. 2.3.1.2
Eidinger, John M. 3.1.17
Engineering News Record. 3.3.1.8
Environmental Cost Handling. 3.3.2.3
Environmental Protection Agency. 4.1.1.17, 2.3.2.3, 2.3.2.2
Environmental Systems Research Institute. Appendix C:5
EQECAT. 3.2.7
ESRI. 2.2.6
Federal Alliance for Safe Homes. 3.1.18
Federal Bureau of Investigation. 2.3.2.6, 2.3.2.5
Federal Emergency Management Agency. Appendix B:7, Appendix B:6, 3.3.1.9, 3.1.34, 3.1.33, 3.1.32, 3.1.31, 3.1.30, 3.1.29, 3.1.28, 3.1.27, 3.1.26, 3.1.25, 3.1.24, 3.1.23, 3.1.22, 3.1.21, 3.1.20, 2.2.8, 2.2.7, 2.1.14, 2.1.13, 2.1.12, 2.1.11, 2.1.10, 2.1.9, 2.1.8, 2.1.7, 2.1.6
Federal Facilities Council. 3.1.35
Federal Highway Administration. 3.3.1.10
Federal Reserve Bank. 4.2.13
Fineberg, Harvey V. Appendix C: 29
Fisk, William J. 3.3.2.4
Frangopol, M. Dan. 3.1.36
Frank R. Walker Company. 3.3.1.11
Fuller, Sieglinde K. 4.1.1.18
Fuller, Sieglinde K.. 4.2.14
Garcia, Mary Lynn. 3.1.37
Godschalk, David R. Appendix C:7
Goebel, R. Matthew. Appendix C:7
Golan, Elise. 3.3.2.7
Government Accountability Office. Appendix C:9, Appendix C:8
Government Printing Office. Appendix B:8
Grance, Tim. 3.1.38
Grossi, Patricia. 3.1.39
Hahn, Robert W. 2.1.16
Hallstrom, Daniel G. Appendix C:28
Hamm, Mark S. Appendix C:10
HardDollar. 3.4.4
Hash, Joan. 3.1.38
Heal, Geoffrey. Appendix C:11
Henderson, LTC Darrall. 4.2.16
Hicks and Associates Incorporated. 3.1.40
High Plains Regional Climate Center and University of Nebraska. 2.3.1.5

Horowitz, John K. 3.3.2.5
Institute for Business and Home Safety. 3.1.41
Integrated Computer Engineering. 2.2.9
Intergraph. 3.2.8
International Association of Earthquake Engineering. 3.1.42, 2.3.1.1
International Code Council. 3.1.43
International Organization of Standardization. 4.1.2.31
International Strategy for Disaster Reduction. 3.1.44
International Trade Administration. 3.3.1.13
International Tsunami Information Centre. 2.3.1.6
Jenicek, Elisabeth M. 3.3.2.8
Jetter, James. Appendix C:12
Joint Chiefs of Staff. 3.1.45
Jones, J. W. 4.2.16
Journal of Homeland Security and Emergency Management. 3.1.46
Kahneman, Daniel. Appendix C:16
Kaiser, Edward J. Appendix C:7
Kettl, Donald F. Appendix C:4
Kuchler, Fred. 3.3.2.7
Kunreuther, Howard. Appendix C:16, Appendix C:15, Appendix C:14, Appendix C:13, Appendix C:11, Appendix C:4, 2.1.17
Layne-Farrar, Anne. 2.1.16
Leite-Monteiro. Appendix C:2
Leng, Chi J. 4.1.1.16, 3.1.47
Leson, Joel. 2.1.18
Lew, H. S. 3.1.13
Lister, Debra Brinegar. 3.3.2.8
Lufkin, Peter S. 3.3.1.14
Mackin, T. J. 4.2.16
Marchand, Maurice. Appendix C:2
Marlatt, Greta E. 2.3.2.8
Marshall, Harold E. Appendix C:17, 4.3.7, 4.1.1.24, 4.1.1.22, 4.1.1.21, 4.1.1.20, 3.1.47
MC^2. 3.4.5
McCarthy, John A. Appendix C:18
McEntire, David A. 3.1.48
Melchers, Robert E. 4.3.9
Meyer, Robert. 2.1.17
Michel-Kerjan, Erwann. Appendix C:14
Michigan Technological University. 2.3.1.7
Mid-America Earthquake Center. 2.3.1.8
Mileti, Dennis. 3.1.49
Military Operations Research Society. 4.2.16
Moore, Michael J. 3.3.2.9
Multidisciplinary Center for Earthquake Engineering Research. 3.1.50
Munich Re. 3.3.2.10
Nadel, Barbara A. 3.1.51
NASA. Appendix B:14, 2.3.1.9
National Academies
    Federal Facilities Council. 3.1.52
National Academy Press. Appendix C:29
National Association of Home Builders Research Center. 3.1.53
National Bureau of Economic Research. Appendix C:14
National Capital Planning Commission. Appendix C:20, Appendix C:19, 3.1.54

National Consortium for the Study of Terrorism. Appendix C:21
National Earthquake Hazards Reduction Program. 3.1.55
National Fire Protection Association. 4.1.2.33, 3.1.57, 3.1.56, 2.3.2.9, 2.1.19
National Incident Management System. 3.1.58
National Information Service for Earthquake Engineering. 3.1.59
National Institute for Occupational Safety. 3.1.60
National Institute of Building Sciences. 3.1.62, 3.1.61
National Memorial Institute for the Prevention of Terrorism. 3.3.2.11, 3.1.40, 2.3.2.11
National Ocean Economics Program. 3.3.2.12
National Research Council. 3.3.2.16, 3.1.67, 3.1.66, 3.1.65
National Science and Technology Council. 3.1.68
National Science Foundation. Appendix C: 24
National Sea Grant Network. 2.3.1.12
National Technology Information Service. 3.3.2.17
National Weather Service. 2.3.1.13
    Office of Climate, Water, and Weather Services. 2.3.1.14
Natural Environment Research Council. 2.3.1.16
Novemsky, Nathan. Appendix C:16
O'Day, Alan. 3.3.2.18
O'Neal Kristofor. 3.1.38
Organization for Economic Co-operation and Dvelopment. Appendix C:25
Owen, David D. 3.3.1.16
Pacific Disaster Center. 2.3.1.17
Pacific Earthquake Engineering Research Center. 3.1.69
Palisade. 2.2.14
Pate-Cornell, M. Elisabeth. Appendix C:26
Pauly, Mark. Appendix C:15
Persily, Andy. 3.1.70
Pestieau, Pierre. Appendix C:2
Petersen, Stephen R. 4.1.1.18
Population Reference Bureau. 3.3.2.19
Porter, Beverly. Appendix C:14
Preissner, Paul Fredrick. 3.3.2.8
Priest, George L. Appendix C:27
Proffitt, David. Appendix C:12
Public Entity Risk Institute. 2.1.20
Quest Solutions. 3.4.6
Ramachandran, Ganapathy. 4.1.1.23
RAND Corporation. 2.1.24, 2.1.21
Reed Construction Data. 3.3.1.16
Risk Management Solutions. Appendix C:38, Appendix C:37, 2.1.26, 2.1.22
Ritz, George J. 3.3.1.17
Robinson, Robie Jack. 3.1.48
Rogers, Mark R. 4.2.21
Rosenfeld, Arthur H. 3.3.2.4
RS Means. 3.4.7, 3.3.1.22, 3.3.1.21, 3.3.1.20, 3.3.1.19, 3.3.1.18, 3.3.1.4
Ruegg, Rosalie T. 4.1.1.24
Sage Timberline Office. 3.4.8
Sandia National Laboratories. 2.2.16
Schiff, Ansel J. 3.1.71
Securac. 2.2.17

Silsbee, Robert M. 3.3.1.14
Smith, V. Kerry. Appendix C:28
Society of fire Protection Engineers. 3.1.57
Stern, Paul C. Appendix C: 29
Stevens, Marc. 3.1.38
Stewart, Mark G. 4.3.9
Swiss Re. 3.3.2.22, 2.3.1.18
Taylor, Craig. 4.1.1.25
Tec-Com Incorporated. 2.2.18
The Infrastructure Security Partnership. 3.1.72
TreeAge. 3.4.9
Tropical Storm Risk Consortium. 2.3.1.19
United States Army Corps of Engineers. 3.1.73
United States Army Training and Doctrine Command. 2.3.2.13
   Deputy Chief of Staff for Intelligence
      Assistant Deputy Chief of Staff for Intelligence. 2.3.2.14
United States Department of Agriculture. 3.3.2.7
United States Department of Commerce
   Bureau of Economic Analysis. 4.2.23
   National Institute of Standards and Technology. Appendix C:23, Appendix C:22, 4.3.7, 4.2.14, 4.1.1.22, 4.1.1.21, 4.1.1.20, 4.1.1.18, 4.1.1.16, 3.1.38, 3.1.13, 2.1.17
      Building and Fire Research Laboratory. 4.1.3.5, 4.1.3.4, 4.1.3.3, 4.1.3.2, 3.1.63, 2.2.10
      Electronics and Electrical Engineering
         Office of Law Enforcement Standards. 3.1.64
      Information Technology Laboratory. 2.2.11
   National Oceanic and Atmospheric Administration. 3.3.2.13, 2.3.1.10
      Coastal and Ocean Resource Economics. 3.3.2.14
      Damage Assessment, Remediation, and Restoration Program. 3.3.2.15
      Satellite and Information Service
         National Environmental Satellite, Data, and Information Service. 2.3.1.15
   United States Census Bureau. 3.3.1.25, 2.3.2.16, 2.3.2.15
United States Department of Defense. 3.1.74
United States Department of Energy. 3.3.1.26
   Energy Efficiency and Renewable Energy. 3.3.1.27
   Energy Information Administration. 3.3.1.32, 3.3.1.31, 3.3.1.30, 3.3.1.29, 3.3.1.28
United States Department of Health and Human Services. Appendix B:24
United States Department of Homeland Security. Appendix C:33, Appendix C:32, Appendix C:31, Appendix C:30, Appendix C:28, Appendix B:26, 3.1.79, 3.1.78, 3.1.77, 3.1.76, 3.1.75
United States Department of Justice. 3.3.2.25
   Bureau of Justice Statistics. 2.3.2.17
United States Department of Labor
   Bureau of Labor Statistics. 4.2.24
United States Department of the Interior
   United States Geological Survey. Appendix C:35, 2.3.1.24, 2.3.1.20
      Earthquakes Hazard Program. 2.3.1.21
      Marine and Coastal Geology Program. 2.3.1.22
      National Geospatial Program Office. 2.3.1.23
      USGS Landslide Hazards Program. 2.3.1.25
      Volcano Hazards Program. 2.3.1.26
United States Department of the Treasury
   Office of Economic Policy. Appendix C:34
United States Department of Transportation. 3.3.1.34
   Bureau of Transportation Statistics. 2.3.2.18
   National Highway Traffic Safety Administration
      National Center for Statistics and Analysis. 3.3.2.27
United States Fire Administration. 2.3.2.19
United States General Services Administration. 3.1.80
United States Government Accountability Office. 3.1.81
University of Colorado-Boulder
   Center for Science and Technology. 2.3.1.27
University of Colorado-Boulder Natural Hazards Center. Appendix B:29
US COST. 3.4.10
Van den Bulte, Christophe. 2.1.17
VanMarcke, Erik. 4.1.1.25
Vertigraph Incorporation. 3.4.11
Viscusi, W. Kip. 3.3.2.29, 3.3.2.28, 3.3.2.9
Weber, Richard T. 3.1.48
White House. Appendix C:36, 3.3.2.30
Whitestone Research. 3.3.1.14
Willis, Henry H. 2.1.24
WinEstimator Incorporated. 3.4.12
Woo, Gordon. Appendix C:38, Appendix C:37, 2.1.26, 2.1.25
World Health Organization. 2.3.1.28
Young, Robert. 3.1.38
Zeckhauser, Richard. Appendix C:39

Subject Index by Reference Number

Adjusted Internal Rate of Return. 4.1.2.17, 4.1.1.22, 4.1.1.18, 4.1.1.16
    Software. 4.1.3.4, 4.1.3.3
Air Quality. 2.3.2.2
    Data and Statistics. 2.3.1.13
    Modeling Software. 2.2.10, 2.2.1
    Productivity. 3.3.2.4
    Reference List. 3.3.2.8
Analytical Hierarchy Process. 4.1.2.13
Annual Loss Expectancy. 2.2.17
Avalanche
    Reference List. Appendix B:14
Benefit-to-Cost Ratio.4.1.2.15, 4.1.1.24
Breakeven Analysis. 4.1.2.10
Bridge
    Design. 3.1.36
    Security. 3.1.11
Building
    Benchmarking Data. 3.3.1.5
    Blast-Effects Mitigation. 3.1.66
    CBR Protection. 3.1.60
    Design and Management Software. 3.2.2
    Earthquake Vulnerability. 2.1.3, 2.1.2
    Engineering Guidelines. 4.1.2.33, 3.3.1.16, 3.1.74, 3.1.62, 3.1.61, 3.1.43, 3.1.20, 3.1.12, 3.1.9, 3.1.8, 2.3.2.9
    Evaluation Methods. 4.1.2.9, 4.1.1.24
    Flood Design. 3.1.7
    Hazard Mitigation. 3.1.47, 3.1.33, 2.1.19
    Security. 3.1.2
    Standards. 4.1.2.2
    Terrorist Insurance. 3.1.30
Business
    Hazard Preparedness. 3.1.79, 3.1.41
Capital Returns. 4.2.21
Catastrophe. *See* Hazard, *and Specific Hazards*
Census Data. 2.3.2.15
Chemical Safety. 2.3.2.1
City Planning. Appendix C:20
    Washington DC Security. 3.1.54
City Planning. Appendix C:19
Climate
    Data and Statistics. 2.3.1.15
    Preparedness. 3.1.18, 2.1.6
    Reference List. Appendix B:29, Appendix B:2
Coastal
    Data and Statistics. 3.3.2.12, 2.3.1.15
    Hazard Mapping. 2.3.1.22
    Hazards. 2.3.1.12
Coefficient of Variation. 4.1.2.10
Cold. *See* Climate
Commercial Building
    Energy Consuption. 3.3.1.31
    Engineering Guidelines. 3.1.31
    Shelter in Place. Appendix C:12
Construction

Bridge Elements. 4.1.2.4
    Cost Analysis. 4.1.2.22
    Cost Classification. 4.1.2.30, 4.1.2.21, 4.1.2.7, 4.1.2.6, 4.1.2.5, 4.1.2.3
    Cost Data. 3.3.1.20, 3.3.1.19, 3.3.1.18, 3.3.1.16, 3.3.1.11, 3.3.1.8, 3.3.1.7
    Cost Engineering. 3.3.1.3, 3.3.1.2
    Cost Estimating Methods. 4.1.2.28, 3.3.1.22, 3.3.1.21, 3.3.1.4
    Cost Estimation Software. 3.4
    Cost Uncertainty. 4.3.7, 4.1.2.16, 4.1.2.10
    Data and Statistics. 3.3.1.25, 2.3.2.16
    Economic Impact. 4.1.2.11
    Engineering Guidelines. 3.1.53
    Industry Outlook. 3.3.1.13
    Industry Statistics. 4.2.23
    Project Management. 3.3.1.17
    Safety Code. 4.1.2.33
    Terms. 4.1.2.24
    Traffic Redirection. 3.3.1.10
    Value Analysis. 4.1.2.23
Consumer Price Index. 4.2.13
Cost Effectiveness. 3.3.2.30
Cost Estimation. 3.3
Cost-Effectiveness Analysis. 3.3.2.7
Cost-of-Illness. 3.3.2.7
Counterterrorism. *See* Terrorism
Crime
    Cyber. Appendix B:2, 2.3.2.14
    Data and Statistics. 3.3.2.25, 2.3.2.17, 2.3.2.5
Cyber Crimes. *See* Crime
Cyclone
    Reference List. Appendix B:14
Dam Failure. 2.1.6
Decision Analysis. 4.1.2.13, 4.1.2.10
Disaster. *See Specific Types*
Discount Rate. 4.2.14, 4.1.1.20, 4.1.1.18, 4.1.1.17
Discounted Payback. *See* Payback
Disease
    Reference List. Appendix B:14
Drought
    Data and Statistics. 3.3.2.13
    Reference List. Appendix B:14, Appendix B:2
Dust Storm
    Data and Statistics. 3.3.2.13
Earthquake
    Data and Statistics. 2.3.1.20, 2.3.1.1
    Engineering Guidelines. 3.1.69, 3.1.59, 3.1.50, 3.1.29, 3.1.15, 2.3.1.8
    Mapping Software. 2.2.7
    Mitigation. 3.1.55, 3.1.21
    Power System Design. 3.1.71
    Preparedness. 3.1.28, 3.1.26, 3.1.25, 3.1.18, 2.1.6
    Reference List. Appendix B:29, Appendix B:14, Appendix B:6, Appendix B:2, 3.1.55, 2.3.1.2
    Vulnerability Assessment. 2.1.3

143

Water Facility Transmission. 3.1.17
Economic
   Building Software. 4.1.3.4, 4.1.3.3
   Data and Statistics. Appendix C:24, 4.2.24, 4.2, 2.3.2.16, 2.3.2.15
   Environmental Sustainability Software. 4.1.3.2
   Evaluation Methods. 4.1.1
   Indicators. 4.2.21, 4.2.13
Emergency
   Response. 3.1.58, 3.1.4, 2.1.6
Employment Rates. 4.2.21, 4.2.13
Energy. 3.3.1.26
   Data and Statistics. 4.2.14, 3.3.1.32, 3.3.1.31, 3.3.1.30, 3.3.1.28
   Efficiency. 3.3.1.27
   Escalation Rate Software. 4.1.3.5
   Outlook. 3.3.1.29
Environment
   Damage Assessment and Restoration. 3.3.2.15
   Economic Building Software. 4.1.3.2
   Economics. 4.1.1.17
   Reference List. 3.3.2.8
   Restoration Cost Data. 3.3.2.3
Event-Related Losses. *See* Hazard Losses
Extreme Temperatures. *See* Climate
Federal Buildings
   Engineering Guidelines. 3.3.2.16, 3.1.65, 3.1.52, 3.1.35, 3.1.13
   Terrorism Mitigation. 3.1.81
Fire
   Data and Statistics. 2.3.2.19, 2.3.2.9
   Disaster Management. 3.1.56
   Engineering Guidelines. 3.1.57, 2.3.2.9
   Mitigation. 4.1.1.23
   Preparedness. 3.1.56, 2.1.6
Flood
   Data and Statistics. 3.3.2.13, 2.1.6, 2.3.1.27, 2.3.1.20, 2.1.6
   Engineering Guidelines. 3.1.7
   Mapping Software. 2.2.7
   Preparedness. 3.1.18, 2.1.6
   Reference List. Appendix B:29, Appendix B:14, Appendix B:2
Fog
   Data and Statistics. 3.3.2.13
Foreign Terrorist Tracking Task Force. Appendix C:3
Geospatial Data and Information. 2.3.1.23
GIS Mapping. 2.2.6
Government
   Preparedness. 3.1.78
   Publications. Appendix B:8
Gross Domestic Product. 4.2.23
   Explained. 4.2.21
Hail
   Data and Statistics. 3.3.2.13
   Preparedness. 3.1.18
Hawthorne Effect. 3.3.2.8
Hazard
   Data and Statistics. Appendix B:1, 3.3.2.22, 3.3.2.10, 2.3.1.28, 2.3.1.14
   Engineering Guidelines. Appendix C:20, Appendix C:19, 3.1.23, 3.1.22

   GIS Data. 3.2.6
   Human Behavior. Appendix C:39
   Losses. 3.3.2
   Mapping Software. 2.2.6
   Maps. 2.1.6
   Mitigation. Appendix B:1, 4.2.16, 3.1.68, 2.1.12, 2.1.11, 2.1.9, 2.1.7
   Modeling. 3.1.39
   Preparedness. Appendix B:3, 3.1.79, 3.1.18, 3.1.3, 2.1.6, 2.1.13, 2.1.6
   Probability Perception. Appendix C:16
   Reference List. Appendix B:26, Appendix B:24, Appendix B:3, 3.1.75, 2.3.1.9
   Response. Appendix C:32, 3.1.64
   Risk Assessment. 2.2.16
   Vulnerability Assessment. 3.2.6, 2.1.10
Hazardous Materials
   Preparedness. 2.2.6
HAZUS. 2.1.6, 2.2.7, 2.1.6
   Guide. 2.1.8
Health
   Data and Statistics. 3.3.2.2, 2.3.2.16
   Value of. 4.1.1.17
Health-Health Analysis. 3.3.2.7
Heat. *See* Climate
High Wind. *See* Wind
Historic Property
   Hazard Mitigation. 2.1.13
Home
   Hazard Preparedness. 3.1.79
   Safety. 3.1.41
Homeland Security. 3.1.20
   Government Capabilities. 3.1.77
   Journal. 3.1.46
   Policies. Appendix C:28
   Reference List. Appendix B:3
Homeland Security Council. Appendix C:3
Homeland Security Digital Library. Appendix B:26
Homeland Security Presidential Directives. Appendix C:3
   Implementation. 3.1.76
Homeland Security Strategic Plan. Appendix C:33
Hospital
   Earthquake Preparedness. 3.1.25
Hurricane
   Data and Statistics. 3.3.2.13, 2.3.1.27
   Katrina. Appendix C:4
   Preparedness. 3.1.18, 2.1.6
   Reference List. Appendix B:29, Appendix B:14, Appendix B:2, 2.3.1.19
Income Data. 2.3.2.15
Industry Standards. 4.1.2
Industry Statistics. 3.3.1.13
Infrastructure
   Design and Management Software. 3.2.8, 3.2.5, 3.2.4, 3.2.3, 3.2.2
   Preparedness. Appendix C:31, Appendix C:18, 3.1.72
Insurance. Appendix C:15, Appendix B:1, 3.3.2.22
   Data and Statistics. 3.3.2.10, 2.3.2.15
   Disaster Loss Mitigation. Appendix C:11
   Public and Private. Appendix C:27

Research. Appendix C:1
	Terrorism. Appendix C:37
Interest Rates. 4.2.13
Internal Rate of Return. 4.1.2.17, 4.1.1.24, 4.1.1.22, 4.1.1.18
Labor
	Data and Statistics. 4.2.24, 3.3.2.1, 3.3.1.18
Landslide
	Data and Statistics. 2.3.1.20
	Mitigation. 2.3.1.25
	Preparedness. 2.1.6
	Reference List. Appendix B:29, Appendix B:14, Appendix B:2
Life-Cycle Cost. 4.1.1.24, 4.1.1.22, 4.1.1.20, 4.1.1.18, 4.1.1.16
	Case Study. 4.1.1.15
	Software. 4.1.3.4, 4.1.3.3
	Standard Practice. 4.1.2.18
Lightning
	Data and Statistics. 3.3.2.13, 2.3.1.27
	Preparedness. 3.1.18
	Reference List. Appendix B:2
Location Cost Adjustment. 3.3.1.18
Man-Made Hazard. 2.3.2
	Mitigation. 2.1.14, 2.1.4
	Toxic Release Data. 2.3.2.3
MasterFormat. 4.1.2.30
Mean-Variance. 4.1.2.10
Mitigation. *See* Individual Hazards
	Costs. 3.3.1
Monte Carlo Simulation. 2.2.14, 2.2.4
Multifamily Housing
	Earthquake Preparedness. 3.1.26
National Flood Insurance Program. 2.1.6
National Icons
	Terrorism Protection. 3.1.81
National Incident Management System. 3.1.58
National Security Strategy. Appendix C:36
Natural Hazard. 2.3.1
	Data and Statistics. 2.3.1.27, 2.3.1.20, 2.3.1.14, 2.3.1.13, 2.3.1.10
	Mapping Software. 2.2.7
	Maps. 2.3.1.17
	Mitigation. Appendix C:7, 3.3.1.9, 3.1.49, 2.1.4
	Reference List. Appendix B:14
	Risk Estimation Software. 2.2.7
Natural Resource Restoration. 3.3.2.15
Net Benefits. 4.1.2.19, 4.1.1.24
Net Savings. 4.1.2.19, 4.1.1.24, 4.1.1.22, 4.1.1.18
	Software. 4.1.3.3
Nuclear
	Preparedness. 2.1.6
Ocupational Health. 4.1.2.1
Office Building
	Engineering Guidelines. 3.1.27
Oil-Spill
	Reference List. Appendix B:2
Payback. 4.1.2.20, 4.1.1.24, 4.1.1.23, 4.1.1.22, 4.1.1.18
Periodicals
	Construction Management Economics. 3.3.1.6
	Journal of Homeland Security and Emergency Management. 3.1.46
	RAND Review. 2.1.21
Population Data. 2.3.2.15
Power Outage
	Preparedness. 3.1.18
Power System
	Earthquake Design. 3.1.71
Preparedness. *See Individual Hazards*
Present Value of Net Savings. 4.1.1.16
	Software. 4.1.3.4
Price Escalation. 4.1.1.18
Producer Price Index. 4.2.13
	Explained. 4.2.21
Productivity
	Indoor Environments. 3.3.2.8, 3.3.2.4
Public Building
	Standards. 4.1.2.34, 3.1.80
Quantity Adjusted Value of Life. 3.3.2.9
RAMCAP. 2.1.4
RAMPART. 2.2.16
Reference List. *See Individual Topics*
Reinsurance. *See Insurance*
Retail Building
	Earthquake Preparedness. 3.1.28
Return on Investment. 2.2.17
Risk. Appendix C:29, 3.3.2.29, 2.1.6
	Analysis Software. 2.2.2
	Assessment. 3.1.20, 2.2.9, 2.2.8, 2.1.17, 2.1.5, 2
	Attitudes. 4.1.1.21
	Data and Statistics. 2.1.20
	Engineering Guidelines. 4.3.9
	Management. Appendix C:26, Appendix B:1, 3, 2.1.20
	Measurements. 2.1.22
	Mitigation. 4.1.2.9, 3.1.44, 2.1.7
	Software. 3.2, 2.2.18, 2.2
Risk-Adjusted Discounting. 4.1.2.10
Risk-Risk Analysis. 3.3.2.7
Savings to Investment Ratio. 4.1.2.15, 4.1.1.24, 4.1.1.22, 4.1.1.18, 4.1.1.16
	Software. 4.1.3.4, 4.1.3.3
School
	Earthquake Rehabilitation. 3.1.29
	Engineering Guidelines. 3.1.32, 3.1.23
Security
	Assessment. 2.1.1
	Educational Material. 3.1.5
	Engineering Guidelines. 3.1.51, 3.1.37, 3.1.10
	Implementation. 3.1.63, 3.1.14
	Information Technology. 3.1.38
	Investments. Appendix C:11
Seismic. *See* Earthquake
Sensitivity Analysis. 4.1.2.10
Severe Weather
	Reference List. Appendix B:29, Appendix B:2
Simple Payback. *See* Payback
Snow
	Data and Statistics. 3.3.2.13
	Reference List. Appendix B:29
Social Costs. 4.1.1.17
Social Insurance. Appendix C:2

Software. 4.1.3, 3.4, 3.2
Software Security. 2.1.16
State Disaster Agencies. Appendix B:2
Terrorism
    Data and Statistics. 3.3.2.25, 3.3.2.11, 2.3.2.11, 2.3.2.6, 2.1.21
    Economic Effects. Appendix C:25
    Engineering Guidelines. 3.1.74, 3.1.33, 3.1.32, 3.1.31, 3.1.20
    Infrastructure Security. 3.1.72
    Insurance. Appendix C:37, Appendix C:34, 3.1.30, 2.1.26
    Mitigation. Appendix C:17, Appendix B:7, 3.1.81, 3.1.66, 3.1.40, 2.1.4
    Preparedness. 3.1.48, 3.1.45, 3.1.18, 2.1.6
    Reference List. Appendix B:3, Appendix B:2, 2.3.2.8
    Research. Appendix C: 21, Appendix C: 10
    Resource Allocation. 4.2.16
    Risk Assessment. 4.1.1.14, 3.1.34, 2.1.25, 2.1.24, 2.1.18, 2.1.5
    Risk Quantification. Appendix C:38
    Tactics. 2.3.2.13
    Tracking Task Force. Appendix C:3
    Trends. 2.1.26
Terrorism Risk Insurance Act. Appendix C:34, Appendix C:14
Thunderstorm
    Data and Statistics. 3.3.2.13
    Preparedness. 3.1.18, 2.1.6
    Reference List. Appendix B:14
Tornado
    Data and Statistics. 3.3.2.13, 2.3.1.27, 2.3.1.18
    Preparedness. 3.1.18, 2.1.6
    Reference List. Appendix B:14
Total Cost of Ownership. 2.2.17
Toxics Release Inventory. 2.3.2.3
Transportation. 3.3.1.34
    Data and Statistics. 3.3.2.27, 2.3.2.18
Tropical Storm. *See* Hurricane
Tsunami
    Data and Statistics. 2.3.1.20, 2.3.1.6
    Preparedness. 3.1.18, 2.1.6, 2.3.1.6, 2.1.6
    Reference List. Appendix B:29, Appendix B:14
    Research. 2.3.1.24, 2.3.1.16
Tunnel
    Security. 3.1.11
Typhoon
    Reference List. Appendix B:14, Appendix B:2
Unemployment Rate. 3.3.2.1
Uniformat II. 4.1.2.21, 4.1.2.6
United States Fire Administration. 2.1.6
United States Government Policy. Appendix C
Universal Task List. 3.1.78
Utility Theory. 4.1.1.23
Value of Life. 4.1.1.23, 3.3.2.29, 3.3.2.28, 3.3.2.9, 3.3.2.7
    Discount Rate. 3.3.2.5
Value of Recreation Sites. 3.3.2.14
Value of Risk. 3.3.2.28
Volcano
    Data and Statistics. 2.3.1.20
    Mitigation. 2.3.1.26
    Reference List. Appendix B:29, Appendix B:14, 2.3.1.7
    Risk. 2.3.1.26
Vulnerability
    Analysis. 4.1.1.25
Vulnerability Assessment Software. 2.2.5, 2.2.3
Water Facility
    Earthquake Evaluation. 3.1.17
Weather
    Data and Statistics. 2.3.1.27, 2.3.1.13
    Reference List. 2.3.1.5
Wildfire
    Data and Statistics. 3.3.2.13, 2.3.1.20
    Preparedness. 3.1.18
    Reference List. Appendix B:29, Appendix B:14
Willingness-to-Pay. 3.3.2.7
Wind
    Mapping Software. 2.2.7
Winter Storm
    Reference List. Appendix B:2
Years to Payback
    Software. 4.1.3.3